T0317580

FUZZY CONTROL AND IDENTIFICATION

FUZZY CONTROL AND IDENTIFICATION

JOHN H. LILLY

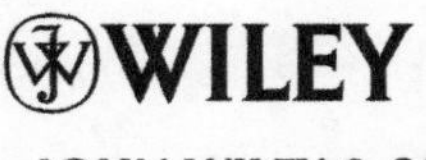

JOHN WILEY & SONS, INC.

Published by John Wiley & Sons, Inc., Hoboken, New Jersey
Published simultaneously in Canada

Library of Congress Cataloging-in-Publication Data:

Lilly, John H., 1949–
Fuzzy control and identification / John H. Lilly.
p. cm.
ISBN 978-0-470-54277-4 (cloth)
1. Fuzzy automata. 2. System identification. 3. Automatic control–Mathematics. I. Title.
TJ213.L438 2010
629.8–dc22

2010007956

10 9 8 7 6 5 4 3 2 1

For Faith, Jack, and Sarah

TABLE OF CONTENTS

PREFACE

In 1982, when I obtained my Ph.D. specializing in adaptive control (the nonfuzzy kind), fuzzy control had not been explored to a very great extent as a research area. There had been only a handful of papers (probably <100) published on the subject up to that time, and some of us "serious researchers" did not take fuzzy seriously as a control method. Since then, of course, the number of papers and books written on some application of fuzzy sytstems has grown to tens of thousands, and many of us "serious researchers," after realizing the potential of the fuzzy approach, have partially or completely redirected our research efforts to some aspect or application of fuzzy identification, classification, or control.

Roughly 10 years after graduating, I started reading anything I could find on the subjects of fuzzy identification and control, culminating in the creation of a graduate-level course on the subject at the University of Louisville. This book is an outgrowth of lectures I presented in this course over the past 10 years, plus some new material that I have not presented yet, but probably will at some point.

I wrote this book to present an introductory-level exposure to two of the principal uses for fuzzy logic: identification and control. This book was written to include topics that I deem important to the subject, but that I could not find all together in any one text. I kept finding myself borrowing material from several sources to teach my course, which is suboptimal for teacher and student alike. In addition, I found that many texts, although excellent, were written on too high a level to be useful as introductory texts. (It is ironic that a subject ridiculed by many as "too easy" quickly becomes so complex as to turn most people away once the basics are covered.) Consequently, I wrote this book, which includes subjects that I think important at hopefully not such a high level as to "blow away" most students.

The book is intended for seniors and first-year graduate students. Some background in control is helpful, but many topics covered in introductory controls courses are of little use here, such as gain and phase margins, root locus, Bode and Nyquist plots, transient and steady-state response, and so on. On the other hand, some of the subjects addressed in this book, such as tracking, model reference, adaptive identification and control, are only covered in advanced-level controls courses. This is in part what makes this subject difficult to teach.

The most helpful preparation would be some understanding of continuous- and discrete-time dynamic systems, and an appreciation of the basic aims and methods of control (i.e., stabilization, tracking, and model reference control). There is little in the way of advanced mathematics beyond differential and difference equations, transfer functions, and linear algebra required to read and understand this book.

The subjects of fuzzy identification and control are quite heavy in computer programming. In order to implement or simulate fuzzy systems, it is almost unavoidable to write computer programs, so it is assumed that the reader is comfortable with at least basic computer programming and computer simulation of dynamic systems. In this book, Matlab is used exclusively for simulations due to its ease of programming matrix manipulations and plotting. I have not relied on any Matlab "canned" programs (e.g., the Matlab differential equation solvers ode23, ode45, etc.) or toolboxes (e.g., the Fuzzy Logic Toolbox). One exception is the use of the *LMI Control Toolbox* used in Chapter 7 to solve a linear matrix inequality. The avoidance of these very powerful specialized tools that Matlab provides was done to give a measure of transparency in the example programs provided in the Appendix, and also because whatever computer language is used to implement these controllers may not (in fact, probably will not) have them.

ARRANGEMENT OF THIS BOOK

The arrangement of this book may seem strange to some. Chapter 5, which presents some well-known nonfuzzy modeling and control methods, may look out of place in the middle of the other chapters, which have to do with only fuzzy topics. It was suggested to me that the material in Chapter 5 either be placed in an introductory chapter or relegated to an appendix. However, I felt there is good reason to place it where it is.

Chapters 2–4 cover basic concepts of fuzzy logic, fuzzy sets, fuzzy systems, and control with Mamdani fuzzy systems. All controllers presented in Chapter 4 are designed on the basis of "expert knowledge." Their design is not based on any mathematical model of the system they control, nor do they use any formal control method (pole placement, tracking, etc.). Therefore, there is no need to study mathematical modeling or control methods to utilize anything through Chapter 4.

On the other hand, Chapters 6 and 7 introduce Takagi–Sugeno (T–S) fuzzy systems, which do necessitate the utilization of a plant model along with choice of some formal control methodology. Thus, the introduction of some standard modeling and control techniques seemed well placed between the Mamdani and T–S developments. I felt that placing this material in either an introductory chapter or an appendix would reduce its chances of being read. At any rate, the chapters are as follows.

Chapter 1 is an introduction to fuzzy logic, fuzzy control, and adaptive fuzzy control. We introduce the concept of *expert knowledge*, which is the basis for much of fuzzy control. We talk briefly about when fuzzy methods may be justified, when they may not, and why. We discuss the plants used in the examples to illustrate various principles taught in this book. Also included in Chapter 1 are brief descriptions of the identification and control problems. Finally, these are combined to discuss the concept of adaptive fuzzy control.

Chapter 2 covers basic concepts of fuzzy sets, such as membership functions, universe of discourse, linguistic variables, linguistic values, support, α-cut, and convexity. We also discuss some set theoretic and logical operations on fuzzy sets, such as fuzzy subset, fuzzy complement, fuzzy intersection, and fuzzy Cartesian product.

Chapter 3 introduces Mandani fuzzy systems, which were historically the first type of fuzzy system used for control. We discuss the various processes that make up fuzzy systems, including fuzzification, inference, defuzzification, and rule base. We discuss the two most common types of defuzzification: center of gravity and center average. In this chapter, we also introduce the concept of the input–output characteristic of a fuzzy controller. Finally, we introduce the singleton fuzzy set, which is used in all subsequent fuzzy identifiers and controllers in this book.

Chapter 4 discusses closed-loop fuzzy control with Mamdani fuzzy systems. It is shown how an effective controller can be designed for many complex nonlinear systems using only common sense. We discuss how the controller can be tuned for improved performance by scaling universes of discourse. We also discuss how fuzzy controllers can be redesigned (again on the basis of common sense) to increase robustness. Chapter 4 includes a method of converting a nonfuzzy proportional-integral-derivative (PID) controller into a fuzzy controller for the purpose of redesigning it to increase robustness. This chapter also includes an introduction to incremental fuzzy control.

Chapter 5 is nonfuzzy. It contains a summary of some common modeling and control techniques that will be used in the rest of the book. It is shown how continuous-time nonlinear systems, which most real-world systems are, can be modeled as fuzzy systems in several forms (continuous-time feedback linearizable form, continuous- and discrete-time linear state-space form, and discrete-time input–output difference equation form). All of these forms are used later in the book. Also included in Chapter 5 are some conventional control methods used in fuzzy control, such as pole placement control, tracking control, model reference control, and feedback linearization. Again, these are introduced because they are used later in the book.

Chapter 6 introduces T–S fuzzy systems as interpolators between memoryless functions, continuous- and discrete-time dynamic systems described in state-space form, and discrete time linear input–output dynamic systems.

Chapter 7 introduces parallel distributed control with T–S fuzzy systems. We introduce the concept of linear matrix inequalities, by which stability can be proved for closed-loop systems involving fuzzy controllers. We discuss how fuzzy tracking and model reference control can be realized for nonlinear systems using parallel distributed controllers.

Chapter 8 discusses the estimation of static nonlinear functions from data using the batch least squares, recursive least squares, and gradient methods. The gradient parameter update equations, similar to *backpropagation* in neural networks, are derived. We address the importance of choice of input data, as well as model validation for these methods.

Chapter 9 uses the principles discussed in Chapter 8 to obtain T–S fuzzy models of dynamic plants for the purpose of using these for closed-loop control. The chapter begins by giving a method of modeling time-invariant nonlinear systems with known mathematical models, with T–S fuzzy systems. The remainder of the chapter is concerned with identification from data of nonlinear continuous time systems as T–S fuzzy systems in either feedback linearizeable or input–output difference equation form.

Chapter 10 uses the principles given in Chapter 9 to develop direct and indirect adaptive fuzzy controllers. These methods are applied to several different systems including a motor-driven robot arm, a ball-and-beam system, and a gantry.

WHAT IS NOT COVERED IN THIS BOOK

A thorough presentation of the topics covered in this book would be quite involved technically and would include a lot of complicated notation. While this is certainly valuable and even indispensable to one wanting to be fully versed in fuzzy control and identification, I have tried to streamline things somewhat in this introductory exposition of the subject by omitting some topics. Specifically, the following topics, which are well known to fuzzy practitioners, are not covered here.

1. ***Fuzzification Methods Other Than Singleton Fuzzification***
   Fuzzification is the method by which measured quantities from real-world systems are converted into fuzzy sets. There are many methods of fuzzification, depending on how the measured quantities are to be interpreted. The most straightforward fuzzification philosophy is *singleton fuzzification*, in which the measured quantities are simply taken to be exact as measured. Since most fuzzification in practice is of the singleton type, this is the only fuzzification strategy considered here.
2. ***Nonsingleton Membership Function Shapes Other Than Triangles and Gaussians***
   In general, the membership function characterizing a fuzzy set should accurately reflect membership in the set. Thus the choice of shapes is virtually unlimited. In practice, however, only a few membership shapes tend to be used much. Of these, we have chosen triangular and Gaussian shapes (triangular because of the possibility of partitions of unity, and Gaussian because of their usefulness in fuzzy identifiers and adaptive controllers).
3. ***Defuzzification Methods Other Than Center of Gravity and Center Average***
   Defuzzification is the method by which fuzzy sets are converted into crisp numbers to be delivered to the outside world. There are numerous defuzzification strategies, but we have chosen to present only the two most common ones: center of gravity and center average defuzzification.
4. ***Overall Implied Fuzzy Sets for Defuzzification***
   Overall implied fuzzy sets are not covered for defuzzification here in favor of the more straightforward individual implied fuzzy sets.
5. ***Universal Approximation Property of Fuzzy Systems***
   The universal approximation property of fuzzy systems states that any function can be approximated by a fuzzy system to arbitrary accuracy. It is important in that it justifies the use of fuzzy identifiers for nonlinear systems. However,

it can be found in many references, and seemed slightly off-topic for the purposes of this book. Therefore, it is omitted.

6. ***Identification Methods Other Than Least Squares and Gradient***
We concentrate on batch and recursive least squares (RLS) and gradient methods and omit other methods, such as orthogonal projection, least-mean squares, clustering, and learning from examples. Also omitted are methods that add momentum to identification techniques.

ACKNOWLEDGMENTS

I would like to acknowledge the help of several people who directly or indirectly influenced me in the writing of this book. First, I have learned much from my Ph.D. students Jerry Branson, Liang Yang, and Jie Liu, all of whom expanded my thinking in fuzzy control in various ways. Second, my colleague Jacek Zurada, who himself is the author of several excellent books on neural networks and computational intelligence, has helped me immeasurably in my career and has given me excellent advice and inspiration. I am grateful to my own Ph.D. advisor, Prof. Mark Balas, for giving me an appreciation of mathematics and rigor in engineering. Finally, I would like to thank Prof. Chi-Tsong Chen, who is the author of numerous excellent textbooks in control, linear systems, and signals, for indirectly inspiring me to write this book. His inspiration came several decades ago when I taught at SUNY Stony Brook, but I have never forgotten it.

CHAPTER 1

INTRODUCTION

The English-language thesaurus has the following synonyms for "fuzzy": ill-defined, indefinite, indistinct, murky, obscure, unclear, vague, and so on. These are words that English-speaking people associate with fuzzy logic, fuzzy control, fuzzy identification, and fuzzy systems. They are not words that anyone would want associated with engineering systems, on which may depend large sums of money, or even worse, peoples' lives. It is unfortunate that the word "fuzzy" was chosen to describe the type of identification and control described in this book. Japanese has no such negative connotations associated with the word "fuzzy," hence systems utilizing fuzzy identification and control are far more prevalent in Japan than in English-speaking countries.

Fuzzy logic, as will be seen in Chapter 3, is modeled on the human reasoning process. Therefore, fuzzy logic is about as "fuzzy" as humans are. A well-designed system utilizing fuzzy logic to perform a task is roughly as dependable at performing the task as a human competent at performing the task would be. Two fuzzy systems designed by different designers to perform the same task may perform it slightly differently, depending on several choices made in the designs. This difference is analogous to the difference that would exist when two different people perform the task, or even the same person on different days. For example, two pilots will land an airplane slightly differently, but each can land it unfailingly every time.

1.1 FUZZY SYSTEMS

When you are driving and choosing which route to take to a desired destination, you usually have several candidate routes from which to choose. You have a set of rules (probably unspoken) in your mind that help you decide which route to take. They might be something like, "If the route distance is short and it does not have many turns, and it does not go on crowded streets, then the route is desirable." Another might be, "If the route goes on narrow streets and it goes close to the area that I just heard on the radio is congested, and it is rush hour, then the route is undesirable." You might have several rules like these in your head, and for any given situation you must somehow balance them all to arrive at the route you will take. This

Fuzzy Control and Identification, By John H. Lilly

decision process is a fuzzy system, and people employ this kind of reasoning all the time, from deciding how to invest their money to deciding which restaurant to go to.

Fuzzy systems are capable of dealing with very complex problems—problems that would be impossible to model mathematically—such as deciding when, where, and how much money to invest. Furthermore, fuzzy systems are employed with success by people with absolutely no technical expertise whatsoever. For instance, a young boy can easily balance an upside-down broom in the palm of his hand for any desired length of time (I remember doing this myself when I was 11). In engineering parlance, this is known as the two-dimensional inverted pendulum problem. This system is difficult to model mathematically, and it is not straightforward to design a controller to keep the broom balanced in the vertical-up position, at least without considerable expertise in control.

If the boy's sister came along and nudged the balanced broom, he could probably regain the balance if the nudge were not too big. Furthermore, if the boy's parakeet, which happened to be flying at large in the house, flew in and landed on the broom while the boy was balancing it, he could probably still balance the broom. Thus the young boy has solved an adaptive control problem for a nonlinear time-varying system subject to disturbances. He has done this with absolutely no expertise in control or math. His reasoning process can be easily and logically expressed in terms of fuzzy systems. The boy might even be able to give a set of linguistic rules he uses to balance the broom.

Fuzzy identification and control methods are used in many engineering systems. Aircraft flight control and navigation systems, which have traditionally used gain scheduling, are now increasingly employing methods of fuzzy control. Some automobile manufacturers use fuzzy logic to control automatic braking systems, transmissions, and suspension systems. In process control systems, fuzzy logic is used to control distillation columns and desalinization processes. In the field of robotics, fuzzy control is used to control end-effector position and path. At least one appliance manufacturer employs a fuzzy system to control turbidity in washing machine water, and at least one camera manufacturer ironically uses fuzzy logic in their autofocus system.

Much is made of the paucity of stability proofs for systems controlled with fuzzy controllers, although inroads have been made with linear matrix inequalities (discussed in Chapter 7). It is true that there exist more stability proofs for closed-loop systems involving conventional nonfuzzy controllers, such as state feedback controllers, H_∞ controllers, sliding mode controllers, adaptive controllers, and the like. However, it must be kept in mind that these proofs all assume some model (or truth model) of the plant being controlled. The true system being controlled is almost never perfectly described by this truth model; the truth model is at best only an approximation of the true plant. This means that the stability of the true plant under state feedback control, H_∞ control, and so on, may not really be guaranteed either.

Finally, we point out that when the greatest precision is needed and money and human life are on the line, automatic control systems, which possess stability proofs, are often eschewed in favor of "fuzzy" humans, which possess no stability proofs. The landing of jumbo jets is seldom entrusted to automatic landing systems; a human pilot usually takes over in the vicinity of the airport. Similarly, the final approach and landing of the Space Shuttle is done by astronauts, not autopilots.

1.2 EXPERT KNOWLEDGE

The term "heuristic" refers to knowledge that is acquired by experimentation or trial and error. Each of us has vast stores of heuristic knowledge that we have accumulated over the years to accomplish many tasks. For instance, you may know how to cook spaghetti or play the piano. These skills did not come instantaneously, they came after much practice. On a much more fundamental level, infants learn to speak, walk, and hundreds of other complicated tasks mainly by practice and trial and error.

It is not uncommon that a person can operate a complex process quite well by him/herself using only heuristic knowledge without the aid of any closed-loop control. For instance, an experienced truck driver can back up a semitrailer to a loading dock without any control system helping him. Pilots can land aircraft using only their experience of past landings. Emergency responders can choose the best routes to take through crowded urban areas from years of experience driving in the area. An experienced investment analyst can know which investments have higher or lower probability for success based on past experience.

These are examples of "expert knowledge," and it is one of the great strengths of fuzzy control to be able to incorporate such knowledge. Incidentally, the term "expert knowledge" is used quite loosely here. For instance, in the above example of a young boy balancing an inverted broom in the palm of his hand, the boy is the "expert."

Some of the controllers in this book are designed using only expert knowledge. In fact, the first fuzzy controllers were designed using only expert knowledge. Some systems that are too complex to permit analytical model-based controller design are easily controlled with fuzzy controllers designed from expert knowledge.

1.3 WHEN AND WHEN NOT TO USE FUZZY CONTROL

Because we usually deal with real-world systems with real-world constraints (cost, computer resources, size, weight, power, heat dissipation, etc.), it goes without saying that the simplest method to accomplish a task is the one that should be used. PID controllers are used in the vast majority of all industrial controllers. They are simple and cheap to construct from discrete components and are quite effective for many control tasks. On the other hand, fuzzy control is usually fairly complex to implement. Therefore, if a control task can be accomplished with a PID or some equally simple controller, that controller should be used instead of a fuzzy one. If the system to be controlled is linear and time invariant, there are many well-known methods for its control; a fuzzy controller for such a system would be overkill.

The strength of the fuzzy approach is in dealing with complex nonlinear systems with perhaps unknown or poorly known mathematical models. For example, the problem of vehicle routing is not easily handled with standard model-based methods, nor is control of distillation columns or control of power systems. Such complex, nonlinear, time-varying, or unknown or poorly known infinite-dimensional systems, while not amenable to analtical methods, can sometimes be handled using fuzzy methods.

1.4 CONTROL

Control is the discipline of forcing a plant to behave as desired [1–3]. The three control objectives addressed in this book are stabilization, tracking, and model following. In stabilization, the control objective is to add or enhance stability. In tracking, the objective is to force the plant output to track a desired reference signal. In model following, the control objective is to force the plant to emulate a reference model that possesses certain qualities desired for the plant.

For an example of stabilization, consider the gantry shown in Figure 1.1.

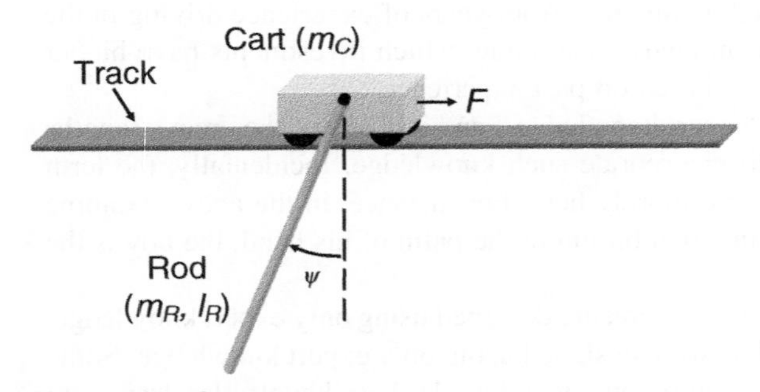

Figure 1.1. Gantry.

The gantry consists of a motorized cart on which is mounted a rigid rod that is free to rotate without friction. The cart moves in one dimension along a track. A typical gantry control objective is to move the cart from location to location along the track with a minimum of rod sway. A practical application of this is found in industry. Industrial gantries are used to move heavy objects from place to place in a building or yard so that they can be worked on or serviced by different machines. It is generally desired that the load sway as little as possible during the movement.

The gantry is open-loop stable since if the rod is displaced from the vertical-down position it will eventually return to it. However, this may take an unacceptably long time because the gantry has very little damping. Therefore, even though the gantry is theoretically stable as it is, we can design a controller to increase the gantry's stability, that is, increase the damping so that oscillations die out more quickly. Large industrial gantries usually use either some type of closed-loop control or an operator with expert knowledge of how to minimize the oscillations.

For another example of stabilization, consider the inverted pendulum of Figure 1.2.

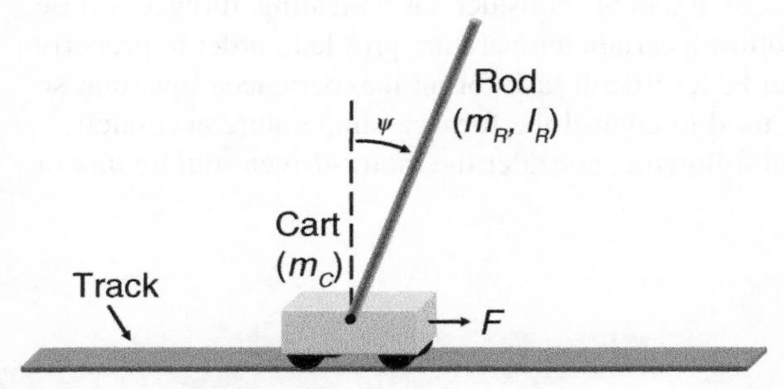

Figure 1.2. Inverted pendulum.

The inverted pendulum consists of a motorized cart on which is mounted a rigid rod that is free to rotate without friction. The cart moves in one dimension along a track.

The inverted pendulum is open-loop unstable (i.e., in the absence of external control), if the rod is displaced from the vertical-up position it will fall down and never return to vertical-up. Some type of closed-loop control is necessary to maintain the rod in the vertical-up position. The stabilization of the inverted pendulum with a fuzzy controller is addressed in Chapter 4.

A practical application of the inverted pendulum is found in rocketry. A rocket immediately after launch, being long and slim, tends to fall over without some type of closed-loop control to keep it vertical. In large modern rockets, there are several subsystems to accomplish this. One is a closed-loop control system to actuate the base of the rocket so that it always moves underneath the rocket's nose. This is essentially a two- dimensional version of the inverted pendulum problem.

For an example of tracking, consider the ball and beam system of Figure 1.3.

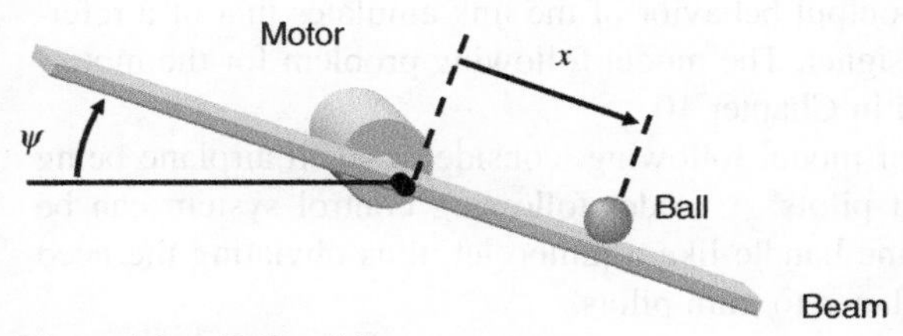

Figure 1.3. Ball and beam system.

In this system, the shaft of a direct current (DC) motor is attached to the center of a beam along which a ball can roll. When power is supplied to the motor, the beam rotates in the vertical plane, causing the ball to roll along it.

The ball and beam is open-loop unstable (i.e., if the ball is displaced from its initial position, it will not return to it without some type of closed-loop control). A tracking control objective would be to actuate the motor so the ball follows a predetermined path at a desired velocity along the beam. The tracking problem for the ball and beam is addressed in Chapter 10.

For a practical example of tracking, consider an annealing furnace whose temperature must accurately follow a certain temperature profile in order to properly anneal certain metals. This can be a difficult task for an inexperienced operator, so closed-loop control is usually used to control the furnace temperature accurately.

For an example of model following, consider the motor-driven robotic link of Figure 1.4.

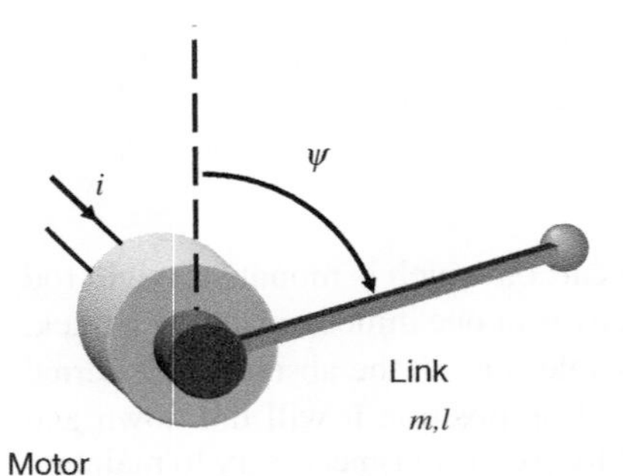

Figure 1.4. Robotic link.

This consists of a DC motor with a rigid rod attached to its shaft. The motor shaft (hence the link) can be rotated through 360°. A true industrial robot might consist of several of these links connected in series (with a second motor and link located at the end of the first link, etc.).

The system is open-loop unstable, that is, if the link is displaced from its initial angle, it will not return to it without some type of closed-loop control (unless the initial angle was vertical-down). A model following control objective would be to actuate the motor so the input–output behavior of the link emulates that of a reference model specified by the designer. The model following problem for the motor-driven robotic link is addressed in Chapter 10.

For a practical example of model following, consider a small airplane being used as a trainer for jumbo jet pilots. A model following control system can be designed to make the small plane handle like a jumbo jet, thus obviating the need to use large and expensive airplanes to train pilots.

1.5 INTERCONNECTION OF SEVERAL SUBSYSTEMS

The identification and control schemes in this book involve interconnections of several subsystems. The best way to describe these is to draw a picture called a *block diagram* showing blocks for each subsystem with lines labeled with the names of the signals they contain interconnecting them. The diagrams of Figures 1.1–1.4 are visually descriptive (they show what the systems look like physically), but not very efficient for showing interconnections between systems.

For instance, consider the gantry of Figure 1.1. Its single input is the force F delivered to the cart, and its output is the rod angle ψ. To depict an interconnection involving the gantry and other subsystems, it is not important what the gantry looks like physically, only its inputs and outputs. Therefore, if the gantry is to be depicted in an interconnection of subsystems, it would be more efficiently represented by the block shown in Figure 1.5. This block shows all pertinent quantities for interconnection, that is the gantry input $F(t)$ and the output $\psi(t)$.

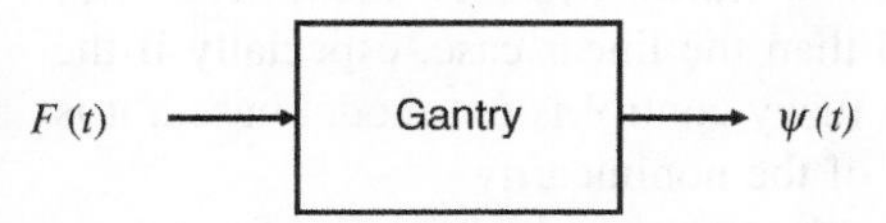

Figure 1.5. Block diagram of gantry.

If we desire to precisely control the gantry, it generally must be placed in some type of feedback configuration. One of the two most basic feedback configurations is the cascade connection with unity feedback. The gantry in a unity feedback configuration is shown in Figure 1.6.

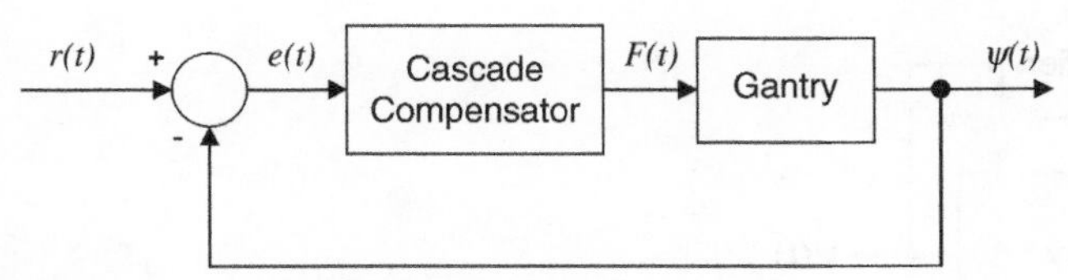

Figure 1.6. Cascade configuration.

Figure 1.6 shows the gantry with its output $\psi(t)$ measured and fed back with negative polarity to one input of a summer. Because ψ is the rod angle, to measure it requires a sensor that can measure angles, perhaps a potentiometer, encoder, resolver, etc. attached to the gantry. The signal actually fed back to the summer is a voltage from the angle sensor that is indicative of the rod angle.

The other summer input is an external reference signal $r(t)$ that may be supplied by the designer as a signal for the gantry angle to follow. Because we desire $\psi(t)$ to follow $r(t)$, $r(t)$ is sometimes called the *command or reference input*. If it is desired that the gantry rod hang vertically downward and motionless, $r(t)$ would be zero.

The summer output is the difference $e(t) = r(t) - \psi(t)$. It is the *error* between the command $r(t)$ and the gantry angle $\psi(t)$. The tracking error $e(t)$ forms the input to a cascade compensator designed to minimize e. The compensator output is a voltage proportional to the force that is to be delivered to the gantry. This voltage is delivered to the cart motor that applies the prescribed force to the gantry. If the compensator is properly designed, this closed-loop system will result in $\psi(t) \rightarrow r(t)$

as $t \to \infty$ (i.e., asymptotic tracking). Since in this book the controllers are fuzzy, the cascade compensator block in Figure 1.6 will be a fuzzy system.

1.6 IDENTIFICATION AND ADAPTIVE CONTROL

In the context of control, identification refers to the determination of a plant model that is sufficient to enable the design of a controller for the plant. Identification of linear time invariant systems is straightforwardly done using conventional methods [4], hence fuzzy identification techniques are not necessary for such systems. Fuzzy techniques are useful for the identification of nonlinear systems. Identification of nonlinear systems is much less well defined than the linear case, especially if the form of the nonlinearity is not known. Since fuzzy control is not model based, it is not necessary to assume any particular form of the nonlinearity.

The identifier takes measurements of the plant input and output, and from these determines a model for the plant. The fact that the plant is nonlinear is not a problem if the identifier is fuzzy because fuzzy systems are in general nonlinear. A block diagram depicting fuzzy identification of the gantry is shown in Figure 1.7.

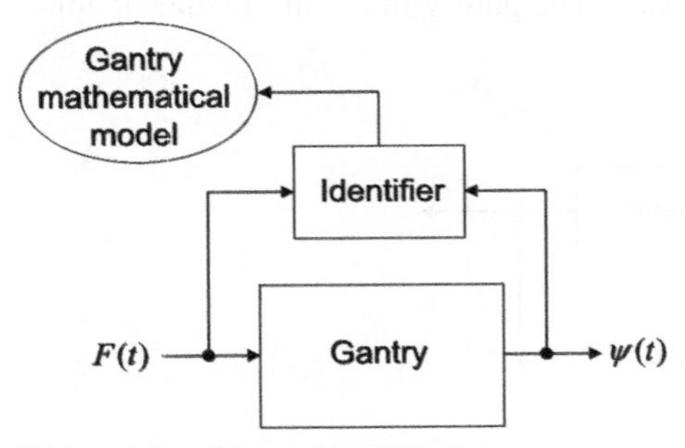

Figure 1.7. Gantry identification.

In Figure 1.7, the force delivered to the gantry (or perhaps compensator output voltage) is measured, as is the gantry rod angle. These two signals, which are voltages, are fed to the identifier, which operates on them to obtain a mathematical model of the gantry. Identification is addressed in Chapter 9.

The latter chapters of this book are concerned with adaptive fuzzy control. Adaptive control is a method by which the system behavior is monitored online in real time and the control continually updated and adjusted to adapt to uncertainties or changes in the plant. There are two basic types of adaptive control: indirect and direct. In indirect adaptive control, the plant is continually identified online, and at each time step during the process, the controller is adjusted based on this identification. This situation is depicted in Figure 1.8. In Figure 1.8, the arrow from the identifier going through the compensator indicates that the compensator is being adjusted in real time by the current mathematical model of the gantry determined by the identifier. In direct adaptive control, which is not depicted here, the parameters of the controller are directly adjusted rather than going through the intermediate step of identification. Direct and indirect adaptive control are addressed in Chapter 10.

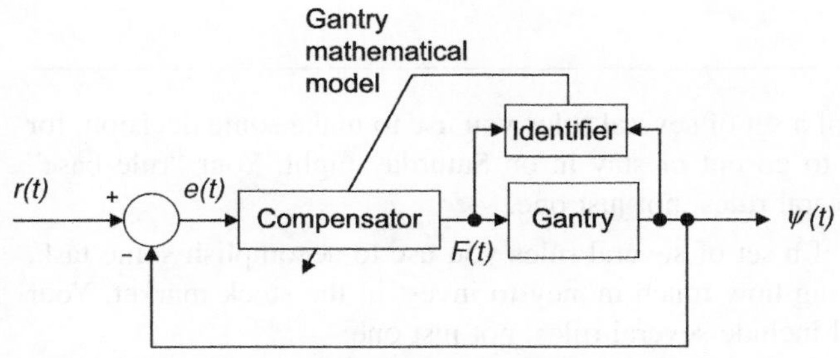

Figure 1.8. Indirect adaptive control of gantry.

Note that block diagrams are not used in the remainder of this book. They are only mentioned here to give the reader some idea of how the various systems in the examples are interconnected for identification and control.

1.7 SUMMARY

Fuzzy logic is an attempt to mimic the human reasoning process. Fuzzy logic can be used to identify and control complicated systems that would be difficult or impossible to control by any other means. **Expert knowledge** is knowledge possessed by human experts about a situation or problem. Expert knowledge, although invaluable in solving complicated problems, cannot be utilized by conventional model-based controllers. However, it is one of the great strengths of fuzzy control that expert knowledge can be easily incorporated into fuzzy controllers.

Fuzzy identification and control is computer-intensive. Therefore, fuzzy methods should be used only when simpler and cheaper methods, such as PID control, cannot accomplish the control task. Fuzzy methods should generally not be used to identify or control linear time invariant systems, as there are many well-established methods for doing this.

The three control objectives addressed in this book are **stabilization**, **tracking**, and **model following**. A system is stable if its response to bounded inputs is bounded, or if it returns to an equilibrium state if displaced from it. The stabilization control objective could involve stabilizing an unstable system, or increasing the stability of an insufficiently stable system. The tracking control objective entails forcing a system's output to track a given reference signal. The model following control objective entails forcing a system to emulate a reference model specified by the designer. The stabilization, tracking, and model following control objectives will be accomplished in this book with fuzzy identifiers and controllers.

Block diagrams are introduced in this chapter only to provide a concise illustration of the way the fuzzy identifiers and controllers in this book will be interconnected with various example systems. The rest of this book does not contain block diagrams.

EXERCISES

1.1 Give an example of a set of several rules you use to make some decision, for instance, whether to go out or stay in on Saturday night. Your "rule base" should include several rules, not just one.

1.2 Give an example of a set of several rules you use to accomplish some task, for instance deciding how much money to invest in the stock market. Your "rule base" should include several rules, not just one.

1.3 Give five examples of expert knowledge that you possess.

1.4 Give five examples of expert knowledge possessed by someone else of whom you are aware.

1.5 Give two examples of systems controlled by fuzzy logic controllers other than the examples given above. State why fuzzy control is appropriate for their control, and why conventional nonfuzzy control would be inadequate.

1.6 Give two examples of systems controlled by conventional nonfuzzy controllers, and state why fuzzy control would be inappropriate for their control.

1.7 Draw block diagrams of the inverted pendulum, the motor driven robotic link, and the ball and beam.

1.8 Draw a block diagram of the inverted pendulum in a unity feedback configuration with cascade compensator.

1.9 Draw a block diagram of the motor driven robotic link with an attached identifier.

1.10 Draw a block diagram of the ball and beam in a unity feedback configuration with an adjustable cascade compensator being adjusted by an identifier.

CHAPTER 2

BASIC CONCEPTS OF FUZZY SETS

This book shows how fuzzy logic can be used for identification and control of dynamic systems. The foundation of fuzzy logic is the fuzzy set. The concept of the fuzzy set was first introduced by Zadeh in [5,6]. The fuzzy set is a generalization of the conventional, or *crisp*, set that is well known to math and engineering students (see, however, even a generalization of the fuzzy set, given in [7]). In this chapter, we give some basic concepts of fuzzy sets that will be useful for the topics covered in this book (i.e., fuzzy sets, universes of discourse, linguistic variables, linguistic values, membership functions, and some associated set-theoretic operations involving them).

2.1 FUZZY SETS

For the purposes of this book, a fuzzy set is a collection of *real numbers* having *partial membership* in the set. This is in contrast with conventional, or *crisp* sets, to which a number can belong or not belong, but not partially belong. For example, consider the set of "all heights of people 6-ft tall or taller." This is a collection of all real numbers ≥6. It is a crisp set because a number either belongs to this set or does not belong to it. It is impossible for a number to partially belong to the set. Now consider a different kind of set, the set of "heights of tall people." A height of 7-ft tall is definitely considered tall, a height of 5-ft tall is definitely not considered tall, and a height of 6-ft tall may be considered "kind of tall," or tall to a certain extent. Because numbers between five and seven can belong to the set with various certainties, the set of "heights of tall people" is a fuzzy set.

As seen above, it takes two things to specify a fuzzy set: the members of the set and each member's degree of membership in the set. Total membership in the set is specified by a membership value of 1, absolute exclusion from the set is specified by a membership value of 0, and partial membership in the set is specified by a membership value between 0 and 1.

If the physical quantity under consideration is described by a word (in the above case, "height") rather than a symbol (say, h), that word is called a *linguistic*

Fuzzy Control and Identification, By John H. Lilly

variable. Fuzzy sets are also usually given linguistic names, called *linguistic values.* For instance, for the "height" variable, we could define a fuzzy set with linguistic value "tall." We could also define fuzzy sets with linguistic values "short" and "medium." Linguistic names are used for variables and their values in fuzzy logic because people usually think and speak in linguistic terms, not mathematical symbols.

The collection of numbers on which a variable is defined is called the *universe of discourse* for the variable. For our purposes, this will usually be the field of real numbers ($\Re$). Often, though, there is an *effective* universe of discourse, which is a finite sized strip of the real line (i.e., $[\alpha, \beta]$, where $\alpha < \beta$ and both are finite). For instance, if we are considering the linguistic variable *height*, the universe of discourse would be the interval $(0, \infty)$ because all heights would fall in this range. However, if we are considering the heights of people, realistically the heights of people range between about 1 and 3 m. Therefore, the effective universe of discourse for "height" would be the interval [1, 3] m.

More formally, consider a variable with universe of discourse $\mathcal{X} \subseteq \Re$, and let x be a real number (i.e., $x \in \mathcal{X}$). Let M denote a fuzzy set defined on $\mathcal{X}$. A *membership function* $\mu^M(x)$ associated with M is a function that maps $\mathcal{X}$ into [0, 1] and gives the degree of membership of $\mathcal{X}$ is M. We say that the fuzzy set M *is characterized by* μ^M. Then the fuzzy set M is defined as

$$M = \{(x, \mu^M(x)) : x \in \mathcal{X}\}$$

This is a pairing of elements from $\mathcal{X}$ with their associated membership values.

For instance, the fuzzy set WARM (a linguistic value), when referring to OUTDOOR TEMPERATURE (a linguistic variable), may be characterized by the membership function of Figure 2.1.

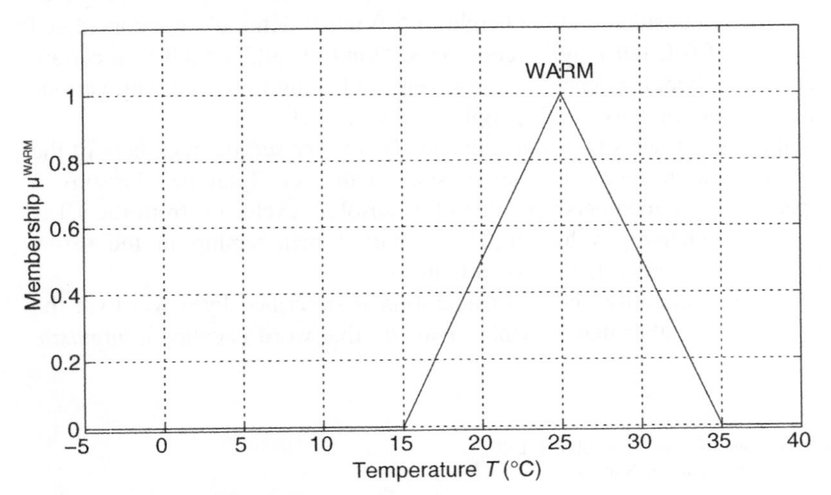

Figure 2.1. Triangular membership function.

This is called a *triangular* membership function, for obvious reasons. It is defined by the conditional function

$$\mu^{\mathrm{WARM}}(T) = \begin{cases} 0.1T - 1.5, & 15 \le T < 25 \\ -0.1T + 3.5, & 25 \le T < 35 \\ 0, & \text{otherwise} \end{cases} \tag{2.1}$$

where $\mu^{\mathrm{WARM}}(T)$ is membership in the WARM fuzzy set and T is temperature.

Note that $\mu^{\mathrm{WARM}}(T)$ is defined for all temperatures T even though it is zero for some T. The universe of discourse for TEMPERATURE is the entire set of possible temperatures $(-273, \infty)$°C, although there may be an effective universe of discourse of, say, $[-20, 50]$°C if it is known that the temperature will never be out of this range.

The membership function indicates that a temperature of 25°C (77°F) is definitely considered warm, temperatures >25°C are decreasingly considered warm as they increase from 25 to 35°C (95°F), and temperatures <25°C are decreasingly considered warm as they decrease from 25 to 15°C (59°F). According to the membership function, temperatures <15°C are not considered warm at all, and temperatures >35°C are also not considered warm at all. A temperature of 20°C is considered warm to a degree 0.5 because $\mu^{\mathrm{WARM}}(20) = 0.5$, and a temperature of 32°C is considered warm to a degree 0.3 because $\mu^{\mathrm{WARM}}(32) = 0.3$.

Another possibility for a membership function to characterize the WARM fuzzy set is the *Gaussian* membership function of Figure 2.2. The mathematical expression for a Gaussian function is

$$\mu(x) = \exp\left(-\frac{1}{2}\left(\frac{x-c}{\sigma}\right)^2\right) \tag{2.2}$$

where c is the center of the function (i.e., the point at which the function attains its maximum of 1, and $\sigma > 0$ determines the spread, or width of the function).

The membership function characterizing the WARM fuzzy set shown in Figure 2.2 conveys similar information to the membership function of Figure 2.1,

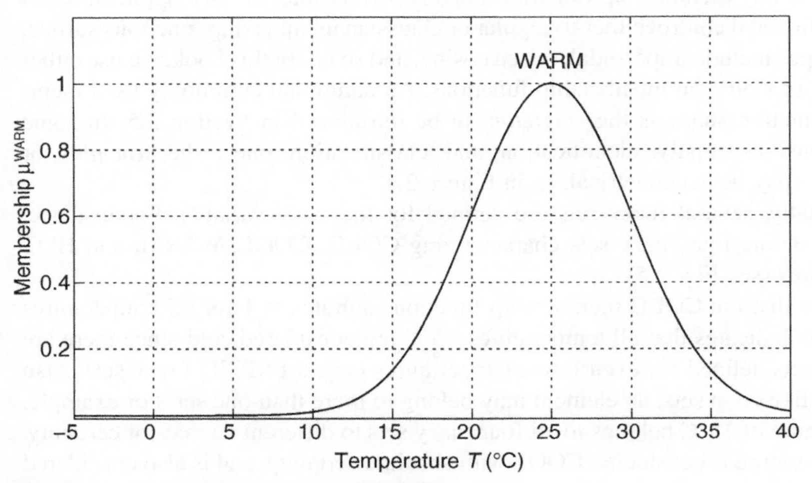

Figure 2.2. Gaussian membership function.

namely, that temperatures close to 25°C are considered warmer, while temperatures further away from 25°C in either direction are considered less warm. The mathematical expression for the membership function in Figure 2.2 is

$$\mu^{\text{WARM}}(T) = \exp\left(-\frac{1}{2}\left(\frac{T-25}{5}\right)^2\right) \tag{2.3}$$

The shape of membership functions is arbitrary. The only requirement is that the membership function make sense for the fuzzy set being defined. For instance, a membership function like that of Figure 2.3 would not make sense if we wanted to characterize the fuzzy set of warm temperatures. From this membership function, it appears that 20 and 30°C are both considered warmer than 25°C!

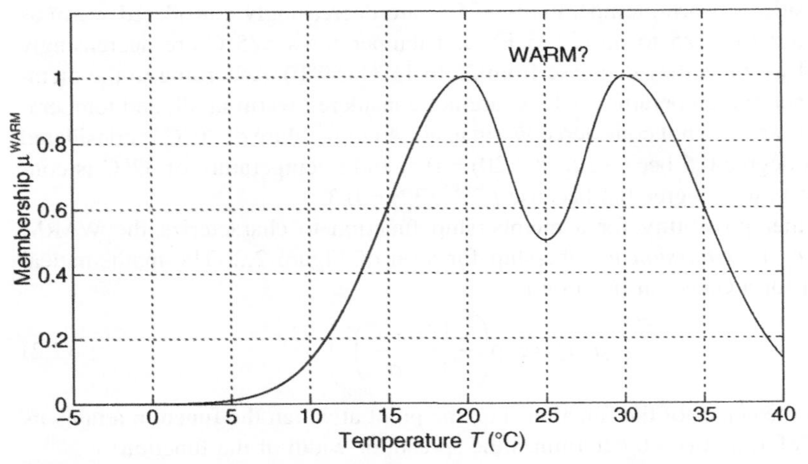

Figure 2.3. Illogical membership function to characterize WARM temperatures.

While any membership function shape is permissible, in most applications of identification and control either triangular or Gaussian membership functions suffice. Other shapes include trapezoidal, raised cosine, and so on. In this book, we use either triangular or Gaussian membership functions. An additional commonly used membership function shape is the *singleton*, to be introduced in Section 2.5. In some applications, especially identification and classification ones, the triangles or Gaussians may be asymmetrical, as in Figure 2.4.

Usually, several fuzzy sets are defined for the same variable. For instance, we could define four fuzzy sets characterizing COLD, COOL, WARM, and HOT temperatures (see Fig. 2.5).

Note that the COLD membership function saturates at 1 for all temperatures <5°C. This indicates that all temperatures <5°C are considered cold since there are no fuzzy sets defined for even lower temperatures (say, a FRIGID fuzzy set). Also note that, like crisp sets, an element may belong to more than one set. For example, a temperature of 15°C belongs to all four fuzzy sets to different degrees of certainty. This temperature is considered COOL with absolute certainty, and is also considered

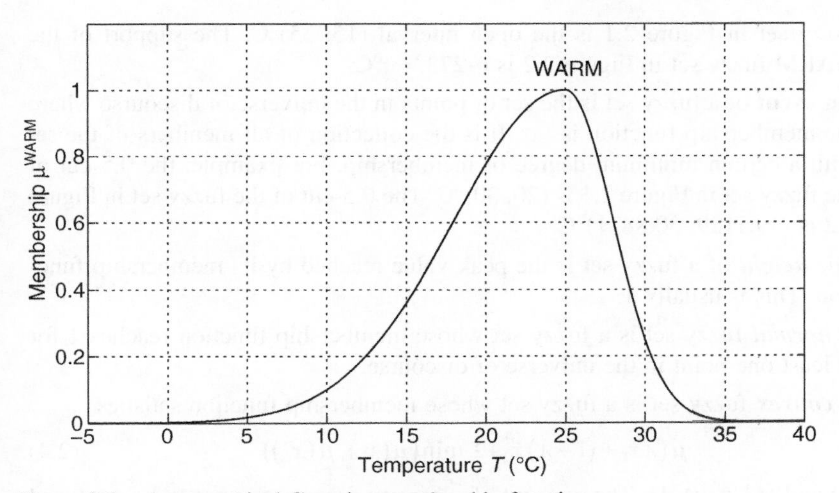

Figure 2.4. Asymmetrical Gaussian membership function.

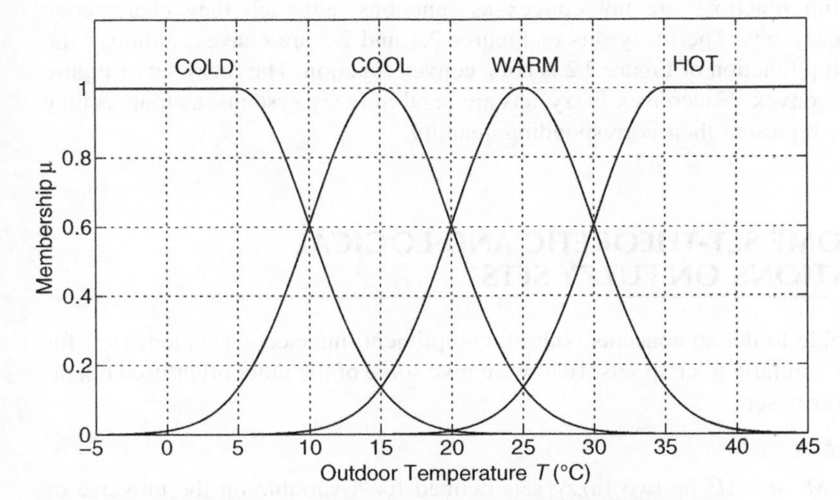

Figure 2.5. Four fuzzy sets defined for the TEMPERATURE variable.

WARM and COLD to lesser extents. This temperature is also considered HOT to a very small extent. In most applications, adjacent fuzzy sets overlap.

2.2 USEFUL CONCEPTS FOR FUZZY SETS

The following are some concepts for fuzzy sets that we will find useful in later studies:

- The ***support*** of a fuzzy set is the set of points in the universe of discourse for which the membership function is >0. For example, the support of the WARM fuzzy set in Figure 2.1 is the open interval (15, 35)°C. The support of the WARM fuzzy set in Figure 2.2 is (–273, ∞)°C.

- An ***α*-cut** of a fuzzy set is the set of points in the universe of discourse where the membership function is $>\alpha$. It is the collection of all members of the set with a certain minimum degree of membership. For example, the 0.5-cut of the fuzzy set in Figure 2.1 is (20, 30)°C. The 0.5-cut of the fuzzy set in Figure 2.2 is (19.1129, 30.8871)°C.
- The ***height*** of a fuzzy set is the peak value reached by its membership function. This is usually 1.
- A ***normal*** fuzzy set is a fuzzy set whose membership function reaches 1 for at least one point in the universe of discourse.
- A ***convex*** fuzzy set is a fuzzy set whose membership function satisfies

$$\mu(\lambda x_1 + (1-\lambda)x_2) \geq \min(\mu(x_1), \mu(x_2)) \tag{2.4}$$

$\forall(x_1, x_2)$ and $\forall\lambda \in (0, 1]$. The concept of a convex fuzzy set is not to be confused with the concept of a convex function, although the definitions look similar. Many membership functions are not convex as functions, although they characterize convex fuzzy sets. The fuzzy sets of Figures 2.1 and 2.2 are convex, although the membership function of Figure 2.2 is not a convex function. The fuzzy set of Figure 2.3 is not convex. Nonconvex fuzzy sets are legal in fuzzy systems, as long as they accurately represent their corresponding quantity.

2.3 SOME SET-THEORETIC AND LOGICAL OPERATIONS ON FUZZY SETS

It is possible to define concepts (subset, compliment, intersection, union, etc.) for fuzzy sets similarly to crisp sets. Below we give some of the more often-used operations on fuzzy sets.

Fuzzy Subset

Let M^1 and M^2 be two fuzzy sets defined for a variable on the universe of discourse $\mathcal{X}$, and let their associated membership functions be $\mu^1(x)$ and $\mu^2(x)$, respectively. Then, M^1 is a *fuzzy subset* of M^2 (or $M^1 \subseteq M^2$) if $\mu^1(x) \leq \mu^2(x) \forall x \in \mathcal{X}$.

Fuzzy Compliment

Consider a fuzzy set M defined for a variable on the universe of discourse $\mathcal{X}$, and let M have associated membership function $\mu^M(x)$. The *fuzzy compliment* of M is a fuzzy set $\overline{M}$ characterized by membership function $\mu^{\overline{M}}(x) = 1 - \mu^M(x)$.

Fuzzy Intersection (AND)

Let M^1 and M^2 be two fuzzy sets defined for a variable on the universe of discourse $\mathcal{X}$, and let their associated membership functions be

$\mu^1(x)$ and $\mu^2(x)$, respectively. The *fuzzy intersection* of M^1 and M^2, denoted by $M^1 \cap M^2$, is a fuzzy set with membership function (1) $\mu^{M^1 \cap M^2}(x) = \min\{\mu^1(x), \mu^2(x) : x \in \mathcal{X}\}$ (*minimum*) or (2) $\mu^{M^1 \cap M^2}(x) = \{\mu^1(x)\mu^2(x) : x \in \mathcal{X}\}$ (*algebraic product*).

Notice that we give two possibilities to characterize the intersection of two fuzzy sets: *min* and *algebraic product.* There are other methods that can be used to represent intersection as well [8–13]. In general, we could use any operation on the two membership functions $\mu^1(x)$ and $\mu^2(x)$ that satisfies the common-sense requirements:

1. An element in the universe cannot belong to the intersection of two fuzzy sets to a greater degree than it belongs to either one of the fuzzy sets individually.
2. If an element does not belong to one of the fuzzy sets, then it cannot belong to the intersection of that fuzzy set and another fuzzy set.
3. If an element belongs to both fuzzy sets with absolute certainty, then it belongs to the intersection of the two fuzzy sets with absolute certainty.

Since the membership function values are between 0 and 1, the operations of *minimum* and *product* both satisfy the above three requirements. These two operations are found to suffice for most fuzzy systems.

If we use the notation * to represent the AND operation, then the membership function characterizing the fuzzy intersection of M^1 and M^2 can be generically written as $\mu^{M^1 \cap M^2}(x) = \mu^1(x) * \mu^2(x)$. The * operation (whether *min, product,* or other) is called a triangular norm or *T-norm.*

Fuzzy Union (Or)

Let M^1 and M^2 be two fuzzy sets defined for a variable on the universe of discourse $\mathcal{X}$, and let their associated membership functions be $\mu^1(x)$ and $\mu^2(x)$, respectively. The *fuzzy union* of M^1 and M^2, denoted by $M^1 \cup M^2$, is a fuzzy set with membership function (1) $\mu^{M^1 \cup M^2}(x) = \max\{\mu^1(x), \mu^2(x) : x \in \mathcal{X}\}$ (*maximum*) or (2) $\mu^{M^1 \cup M^2}(x) = \{\mu^1(x) + \mu^2(x) - \mu^1(x)\mu^2(x) : x \in \mathcal{X}\}$ (*algebraic sum*).

Notice that we give two possibilities to characterize the union of two fuzzy sets: *max* and *algebraic sum.* There are other methods that can be used to represent union as well (see the above references). In general, we could use any operation on the two membership functions $\mu^1(x)$ and $\mu^2(x)$ that satisfies the common-sense requirements:

1′. An element in the universe cannot belong to the union of two fuzzy sets to a lesser degree than it belongs to either one of the fuzzy sets individually.

2′. If an element belongs to one of the fuzzy sets, then it must belong to the union of that fuzzy set and another fuzzy set.

3′. If an element does not belong to either fuzzy set, then it cannot belong to the union of the two fuzzy sets.

Since the membership function values are between 0 and 1, the operations of *maximum* and *algebraic sum* both satisfy requirements 1′–3′. These two operations are found to suffice for most fuzzy systems.

If we use the notation ⊕ to represent the OR operation, then the membership function characterizing the fuzzy union of M^1 and M^2 can be written as $\mu^{M^1 \cup M^2}(x) = \mu^1(x) \oplus \mu^2(x)$. The ⊕ operation (whether *max, algebraic sum*, or other) is called a triangular conorm or *T-conorm.*

Fuzzy Cartesian Product

Cartesian product refers to combining fuzzy sets defined for *different* variables on *different* universes of discourse. If M_1^j, M_2^k, … , M_n^l are fuzzy sets defined on the universes $\mathcal{X}_1$, $\mathcal{X}_2$, … , $\mathcal{X}_n$, respectively, with fuzzy set M_i^m characterized by membership function μ_i^m, their Cartesian product is a fuzzy set, denoted by $M_1^j \times M_2^k \times \cdots \times M_n^l$, characterized by the membership function

$$\mu^{M_1^j \times M_2^K \times \cdots \times M_n^1}(x_1, x_2, \ldots, x_n) = \mu_1^j(x_1) * \mu_2^k(x_2) * \cdots * \mu_1^l(x_n) \tag{2.5}$$

where the T-norm * is defined as above (*min, product*, etc.).

2.4 EXAMPLE

Let the COMFORTABLE ROOM TEMPERATURE (CRT) fuzzy set be characterized by the membership function

$$\mu^{CRT}(T) = \exp\left(-0.5\left(\frac{T-25}{2}\right)^2\right) \tag{2.6}$$

which is shown in Figure 2.6.

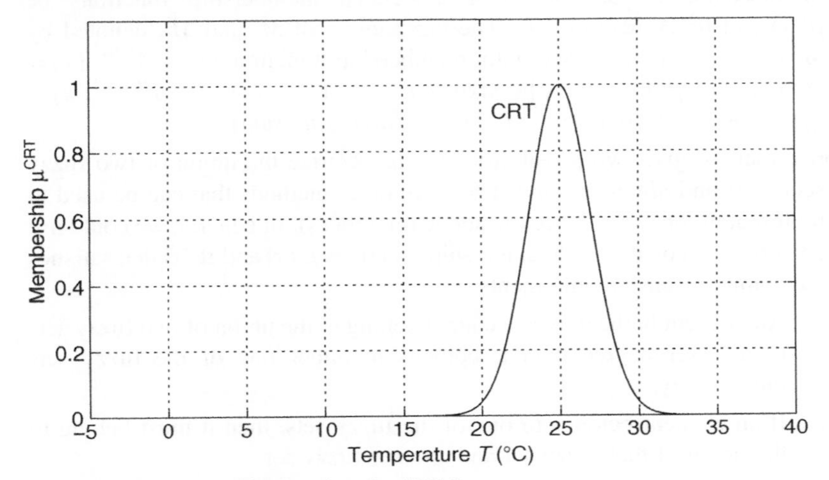

Figure 2.6. Membership function for CRT fuzzy set.

Now it is clear that the CRT fuzzy set is a subset of the WARM fuzzy set shown in Figure 2.2, because $\mu^{\text{CRT}}(T) \leq \mu^{\text{WARM}}(T) \forall T$. A plot of $\mu^{\text{CRT}}(T)$ and $\mu^{\text{WARM}}(T)$ together is shown in Figure 2.7.

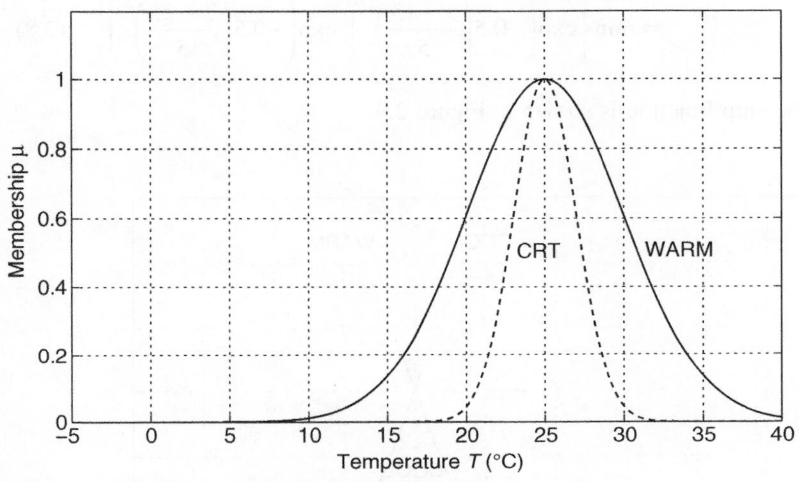

Figure 2.7. Graphical illustration of a fuzzy subset.

The compliment of the WARM fuzzy set (i.e., the NOT WARM) or $\overline{\text{WARM}}$ fuzzy set, is characterized by the membership function

$$\mu^{\overline{\text{WARM}}}(T) = 1 - \exp\left(-0.5\left(\frac{T-25}{5}\right)^2\right) \tag{2.7}$$

and shown in Figure 2.8.

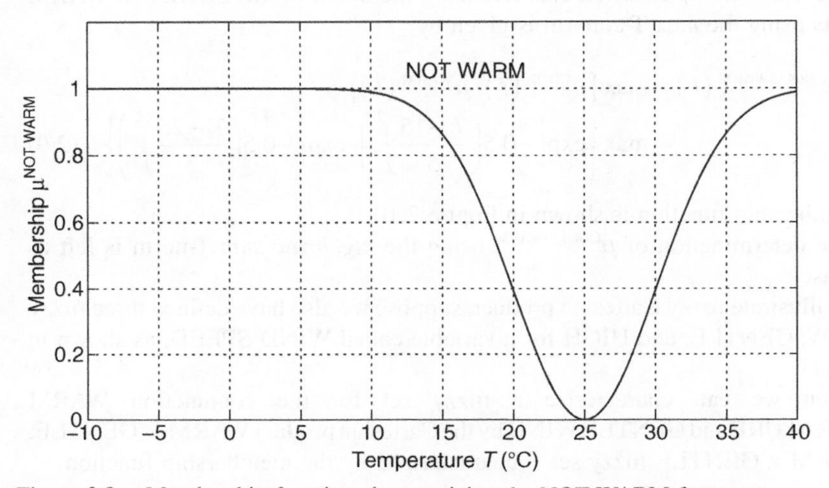

Figure 2.8. Membership function characterizing the NOT WARM fuzzy set.

The membership function characterizing the intersection of the COOL and WARM fuzzy sets using the *min* T-norm is given by

$$\mu^{\mathrm{COOL}\cap\mathrm{WARM}}(T) = \min\{\mu^{COOL}(T), \mu^{WARM}(T)\}$$
$$= \min\left\{\exp\left(-0.5\left(\frac{T-15}{5}\right)^2\right), \exp\left(-0.5\left(\frac{T-25}{5}\right)^2\right)\right\} \quad (2.8)$$

This membership function is shown in Figure 2.9.

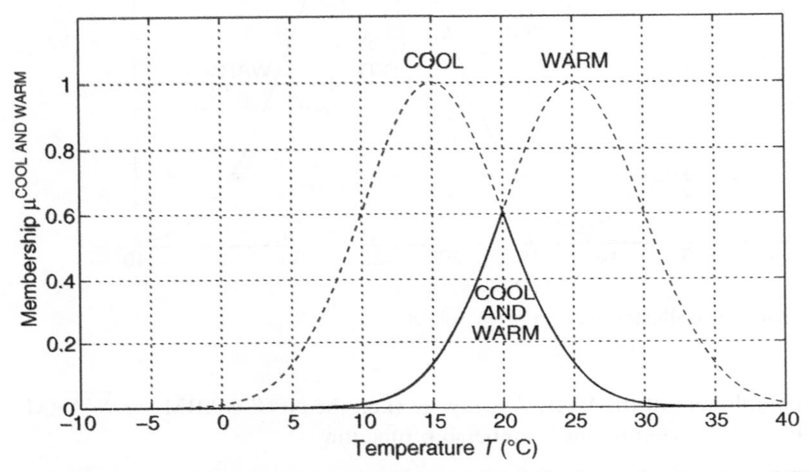

Figure 2.9. The membership function characterizing the fuzzy. intersection of COOL and WARM fuzzy sets.

The determination of μ using the *algebraic product* T-norm is left as an exercise.

The membership function characterizing the union of the COOL and WARM fuzzy sets using the *max* T-conorm is given by

$$\mu^{\mathrm{COOL}\cup\mathrm{WARM}}(T) = \max\{\mu^{\mathrm{COOL}}(T), \mu^{\mathrm{WARM}}(T)\}$$
$$= \max\left\{\exp\left(-0.5\left(\frac{T-15}{5}\right)^2\right), \exp\left(-0.5\left(\frac{T-25}{5}\right)^2\right)\right\} \quad (2.9)$$

This membership function is shown in Figure 2.10.

The determination of $\mu^{\mathrm{COOL}\cup\mathrm{WARM}}$ using the *algebraic sum* T-norm is left as an exercise.

To illustrate fuzzy Cartesian product, suppose we also have defined three fuzzy sets, LOW, GENTLE, and HIGH for a variable called WIND SPEED, as shown in Figure 2.11.

Then we can characterize a fuzzy set for the conjunction WARM TEMPERATURE and GENTLE WIND by the Cartesian product WARM × GENTLE. The WARM × GENTLE fuzzy set is characterized by the membership function

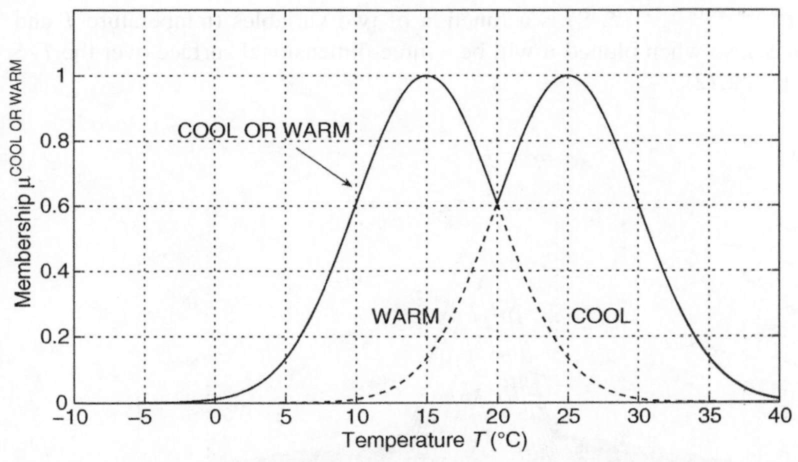

Figure 2.10. The membership function characterizing the fuzzy union of COOL and WARM fuzzy sets.

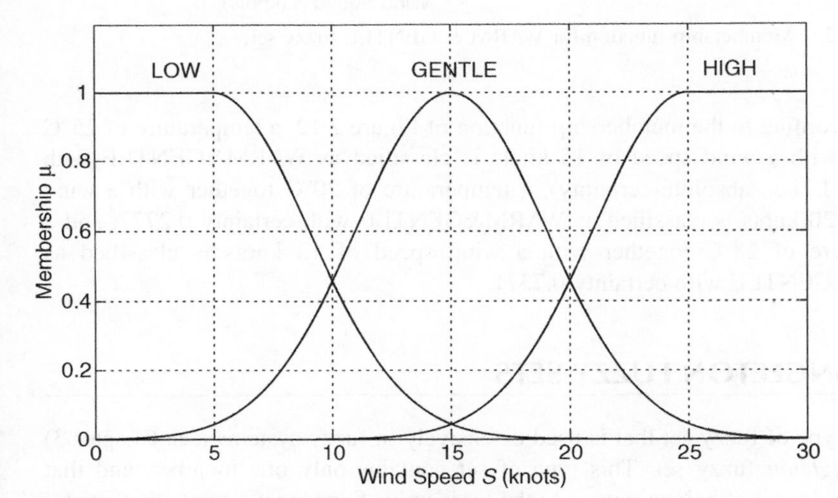

Figure 2.11. Fuzzy sets defined for WIND SPEED.

$$\mu^{\text{WARM and GENTLE}}(T, S) = \mu_{\text{TEMP}}^{\text{WARM}}(T) * \mu_{\text{WIND}}^{\text{GENTLE}}(S) \tag{2.10}$$

where subscripts denote the linguistic variable and superscripts denote the linguistic value.

Using the *product* T-norm, this gives

$$\mu^{\text{WARM and GENTLE}}(T, S) = \exp\left(-0.5\left(\frac{T-25}{5}\right)^2\right)\exp\left(-0.5\left(\frac{S-15}{4}\right)^2\right) \tag{2.11}$$

Note that $\mu^{\text{WARM and GENTLE}}(T, S)$ is a function of two variables (temperature T and wind speed S), so when plotted it will be a three-dimensional surface over the T–S plane (see Fig. 2.12).

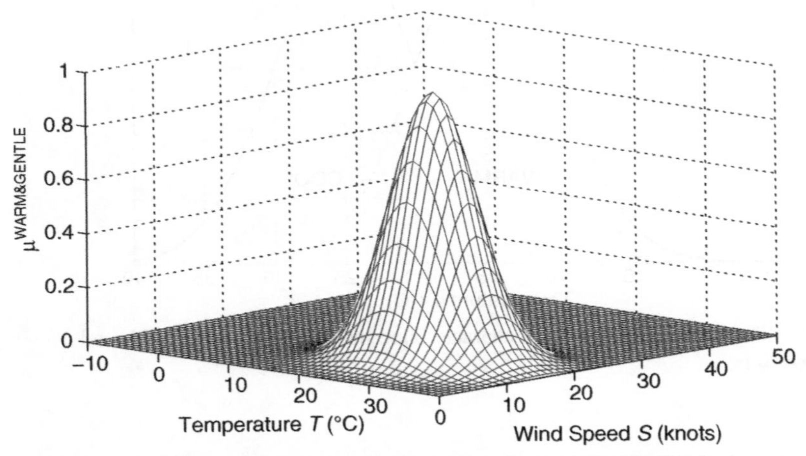

Figure 2.12. Membership function for WARM & GENTLE fuzzy set.

According to the membership function of Figure 2.12, a temperature of 25°C together with a wind speed of 15 knots is classified as WARM&GENTLE with certainty 1 (i.e., absolute certainty), a temperature of 20°C together with a wind speed of 20 knots is classified as WARM&GENTLE with certainty 0.2777, and a temperature of 28°C together with a wind speed of 13 knots is classified as WARM&GENTLE with certainty 0.7371.

2.5 SINGLETON FUZZY SETS

Another type of fuzzy set that is used extensively in fuzzy systems (see Chapter 3) is the *singleton* fuzzy set. This type of set contains only one member, and that member's degree of belongingness to the set is unity. Suppose we create five singleton fuzzy sets on the VOLTAGE universe of discourse such that a NEGATIVE LARGE (NL) voltage is defined as exactly −1 V, a NEGATIVE SMALL (NS) voltage is defined as exactly −0.5 V, a ZERO (Z) voltage is defined as exactly 0 V, a POSITIVE SMALL (PS) voltage is defined as exactly 0.5 V, and a POSITIVE LARGE (PL) voltage is defined as exactly 1 V. We can consider these five voltages as the only members of five singleton fuzzy sets characterized by the membership functions shown in Figure 2.13.

Although it seems unnecessarily complicated to do this, the characterization of certain quantities as members of singleton fuzzy sets is very useful for fuzzification and certain output fuzzy sets (see Chapter 3).

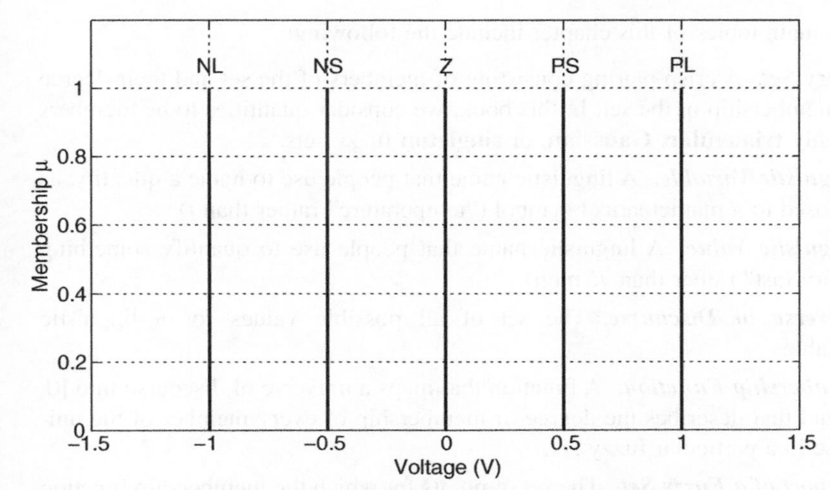

Figure 2.13. Membership functions characterizing five singleton fuzzy sets.

2.6 SUMMARY

Real-world processes, whether they are electrical systems, mechanical systems, heat transfer systems, financial systems, or whatever, involve crisp quantities. When you invest, you must invest an exact quantity of money, say $1000, not "a lot of money." When you make a pizza, you must set the oven to an exact temperature, say 450°F, you do not turn it up until it's "pretty hot."

The fuzzy reasoning process cannot deal with crisp numbers like these. It is modeled after the human reasoning process, which is also not very efficient at dealing with crisp numbers. When we make decisions, we usually make them based on our feelings, perceptions, past experience, and intuition. Our brains are very efficient at deciding things this way. When deciding how much of a tip to leave the waiter, we might say to ourselves, "the service was prompt, so I will leave a nice tip" and be done with it, instead of, "my food was brought in 10.25 min, so my tip will be 15% plus 2.5% times the number of minutes less than the average response time it took to bring my food."

Fuzzy identification and control use fuzzy reasoning processes to make decisions about how to model and/or control real-world systems. Since real-world systems operate in the crisp realm and fuzzy reasoning cannot deal with crisp quantities, fuzzy sets are used as a means of translating between the two. As will be seen, the translation goes both ways. Fuzzy sets are used to convert crisp numbers from the real world into fuzzy quantities that can be utilized by the fuzzy reasoning process. Fuzzy sets are also used to convert the recommendations of the fuzzy reasoning process into crisp numbers that can be utilized by the real-world process.

The main topics of this chapter include the following:

1. ***Fuzzy Set.*** A crisp pairing consisting of members of the set and their degree of membership in the set. In this book, we consider quantities to be members of only **triangular**, **Gaussian**, or **singleton** fuzzy sets.
2. ***Linguistic Variable.*** A linguistic name that people use to name a quantity, as opposed to a mathematical symbol ("temperature" rather than t).
3. ***Linguistic Value.*** A linguistic name that people use to quantify something ("very fast" rather than 75 mph).
4. ***Universe of Discourse.*** The set of all possible values for a linguistic variable.
5. ***Membership Function.*** A function that maps a universe of discourse into [0, 1] and that describes the degree of membership of every member of the universe in a particular fuzzy set.
6. ***Support of a Fuzzy Set.*** The set of points for which the membership function is >0.
7. ***α-Cut of a Fuzzy Set.*** The set of points for which the membership function is $>\alpha$.
8. ***Height of Fuzzy Set.*** The peak value reached by its membership function.
9. ***Normal Fuzzy Set.*** A fuzzy set whose membership function reaches a value of 1 for at least one point in the universe.
10. ***Convex Fuzzy Set.*** A fuzzy set whose membership function μ satisfies $\mu(\lambda x_1 + (1-\lambda)x_2) \geq \min(\mu(x_1), \mu(x_2))\ \forall\ (x_1, x_2)$ and $\forall \lambda \in (0, 1]$.
11. ***T-Norm (Triangular Norm).*** Any operation between fuzzy sets that preserves the three axioms 1–3 in Section 2.3.
12. ***T-Conorm (Triangular Conorm).*** Any operation between fuzzy sets that preserves the three axioms $1'$–$3'$ in Section 2.3.
13. ***Fuzzy Subset.*** A fuzzy set all of whose members also belong to another fuzzy set on the same universe.
14. ***Fuzzy Compliment.*** A fuzzy set whose members also belong to another fuzzy set on the same universe with inverse belongingness.
15. ***Fuzzy Intersection.*** A fuzzy set whose members also belong to several other fuzzy sets that are all on the same universe.
16. ***Fuzzy Union.*** A fuzzy set whose members also belong to any of several other fuzzy sets that are all on the same universe
17. ***Fuzzy Cartesian Product.*** A fuzzy set whose members also belong to several other fuzzy sets that are all on different universes.

EXERCISES

2.1 Give mathematical expressions for $\mu_1(x)$ and $\mu_2(x)$ shown below (Figs. 2.14 and 2.5).

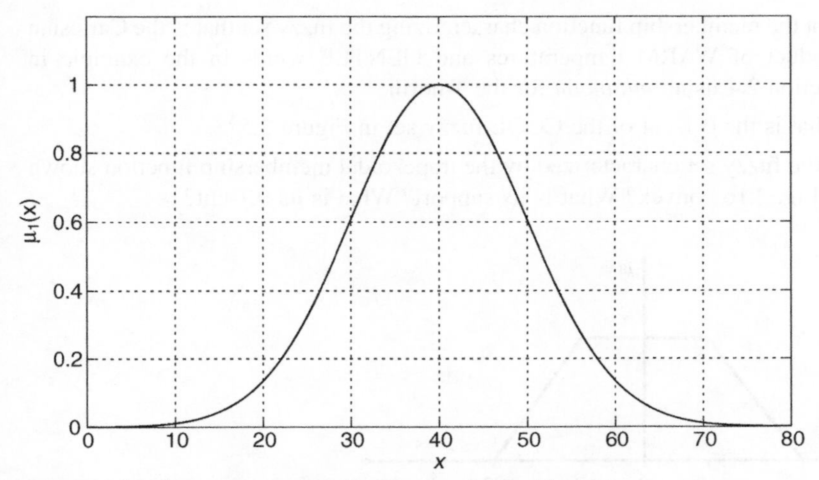

Figure 2.14. Gaussian membership function for Problem 2.1.

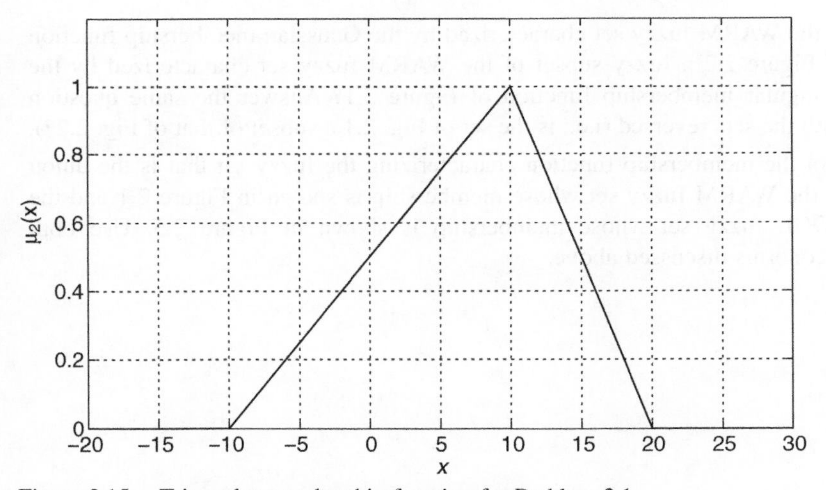

Figure 2.15. Triangular membership function for Problem 2.1.

2.2 Calculate the mathematical expressions for the Gaussian memberships in Figures 2.5 and 2.11.

2.3 (a) Specify a fuzzy set for all "fast" car speeds. (b) Specify five fuzzy sets on the "car speed" universe of discourse.

2.4 Plot the membership function characterizing the COOL ∩ WARM fuzzy set in the example in Section 2.4 using *algebraic product* for the T-norm.

2.5 Plot the membership function characterizing the COOL ∪ WARM fuzzy set in the example in Section 2.4 using *algebraic sum* for the T-norm.

2.6 Plot the membership function characterizing the fuzzy set that is the Cartesian product of WARM temperatures and GENTLE winds in the example in Section 2.4 using *minimum* for the T-norm.

2.7 What is the 0.1-cut of the COOL fuzzy set in Figure 2.5?

2.8 Is the fuzzy set characterized by the trapezoidal membership function shown in Fig. 2.16 convex? What is its support? What is its 0.1-cut?

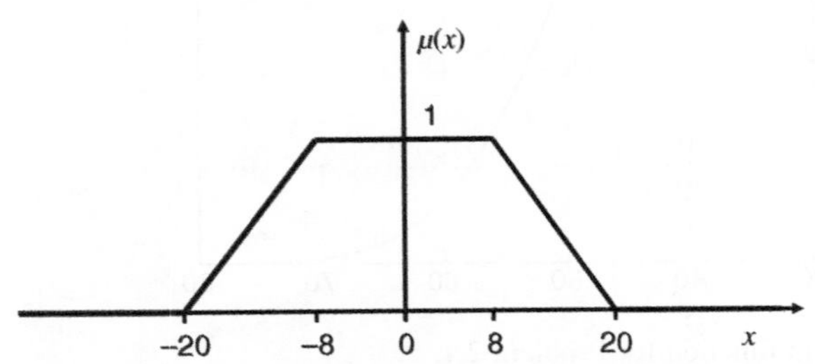

Figure 2.16. Trapezoidal membership function for Problem 2.8.

2.9 Is the WARM fuzzy set characterized by the Gaussian membership function of Figure 2.2 a fuzzy subset of the WARM fuzzy set characterized by the triangular membership function of Figure 2.1? Answer the same question with the sets reversed (i.e., is the set of Fig. 2.1 a subset of that of Fig. 2.2?).

2.10 Plot the membership function characterizing the fuzzy set that is the union of the WARM fuzzy set whose membership is shown in Figure 2.1 and the COOL fuzzy set whose membership is shown in Figure 2.5. Use both T-conorms discussed above.

CHAPTER 3

MAMDANI FUZZY SYSTEMS

A system is a combination of components that as a whole operate on a vector of input functions of time $x(t) \in \Re^n$ for each t to produce a vector of output functions of time $y(t) \in \Re^m$ for each t. A fuzzy system is a system that uses fuzzy logic to operate on the input $x(t)$ to produce the crisp output $y(t)$. It can be shown that a fuzzy system with n inputs and m outputs (i.e., a multi-input multi-output, or MIMO fuzzy system) is equivalent to m fuzzy systems, each with n inputs and one output (i.e., multi-input single-output, or MISO fuzzy systems). Therefore, we will study only MISO fuzzy systems in this book. All fuzzy controllers and identifiers are fuzzy systems.

3.1 IF-THEN RULES AND RULE BASE

Much of human decision making occurs in the form of "if-then" rules. The four forms of if-then rules in classical logic are called *modus ponendo ponens* (Latin for "mode that affirms by affirming") or simply *modus ponens*, *modus tollendo tollens* (Latin for "mode that denies by denying"), or simply *modus tollens*, *modus ponendo tollens* (Latin for "mode that denies by affirming"), and *modus tollendo ponens* (Latin for "mode that affirms by denying").

For examples of these, consider the problem of steering a vehicle toward a target in the presence of an obstacle. Possible rules for deciding which way to steer might be (this is not a complete set):

1. If the target direction is ahead, then the steering direction is straight (modus ponendo ponens).
2. If the obstacle direction is ahead then the steering direction is not straight (modus ponendo tollens).
3. If the obstacle direction is not ahead then the steering direction is straight (modus tollendo ponens).
4. If the target direction is not ahead then the steering direction is not straight (modus tollendo tollens).

Fuzzy Control and Identification, By John H. Lilly

In statement 1, the recommendation to steer straight is affirmed by affirming that the target is ahead. In statement 2, the recommendation to steer straight is denied by affirming that an obstacle is ahead. In statement 3, the recommendation to steer straight is affirmed by denying that the obstacle is ahead. In statement 4, the recommendation to steer straight is denied by denying that the target is ahead. Of course, a complete set of rules for steering to a target in the presence of obstacles would require more rules than the above. People reason in all of these ways. All of these modes of reasoning can be implemented with fuzzy logic (see [14] and [15] for practical examples of controllers employing *modus ponendo tollens* logic).

The vast majority of if-then rules used in fuzzy control and identification are of the *modus ponens* form. For example, consider the rule R_i:

$$R_i \qquad \text{If } \tilde{x} \text{ is } \tilde{P}, \text{ then } \tilde{y} \text{ is } \tilde{Q} \tag{3.1}$$

where $\tilde{x}$ is a linguistic variable defined on universe $\mathcal{X}$, $\tilde{P}$ is a linguistic value described by fuzzy set P defined on universe $\mathcal{X}$, $\tilde{y}$ is a linguistic variable defined on universe $\mathcal{Y}$, and $\tilde{Q}$ is a linguistic value described by fuzzy set Q defined on universe $\mathcal{Y}$. A tilde over a symbol indicates a linguistic variable or value.

The first part of the statement, "$\tilde{x}$ is $\tilde{P}$," is called the *premise* of the rule, and the second part of the statement, "$\tilde{y}$ is $\tilde{Q}$ " is called the *consequent* of the rule. The consequent of the rule is affirmed by affirming the premise—if the premise is true, the consequent is also true.

An example of an if-then rule in *modus ponens* form pertaining to stopping a car is, "If SPEED is FAST then BRAKE PRESSURE is HEAVY." In this rule, SPEED is FAST is the premise and BRAKE PRESSURE is HEAVY is the consequent. SPEED is the input linguistic variable, FAST is a linguistic value of SPEED and is a fuzzy set on the SPEED universe, BRAKE PRESSURE is the output linguistic variable, and HEAVY is a linguistic value of BRAKE PRESSURE and is a fuzzy set on the BRAKE PRESSURE universe.

Note that there may be more than one part to the premise, that is, we could have the rule R_j:

$$R_j \qquad \text{If } \tilde{x}_1 \text{ is } \tilde{P}_1^k \text{ and } \tilde{x}_2 \text{ is } \tilde{P}_2^l \text{ and} \cdots \text{and } \tilde{x}_n \text{ is } \tilde{P}_n^m, \text{ then } \tilde{y} \text{ is } \tilde{Q}^j \tag{3.2}$$

In this rule, the premise is a conjunction of n conditions: $\tilde{x}_1$ is $\tilde{P}_1^k$ and $\tilde{x}_2$ is $\tilde{P}_2^l$ and ... and $\tilde{x}_n$ is $\tilde{P}_n^m$. For example, another rule about stopping a car might be, "If SPEED is FAST and GRADE is DOWNHILL then BRAKE PRESSURE is VERY HEAVY."

In general, we use a number of rules to accomplish a task. Several rules are needed to specify actions to be taken under different conditions. The collection of rules as a whole is called a *Rule base*. A simple rule base for stopping a car might be

1. If SPEED is SLOW, then BRAKE PRESSURE is LIGHT.
2. If SPEED is MEDIUM, then BRAKE PRESSURE is MEDIUM.
3. If SPEED is FAST, then BRAKE PRESSURE is HEAVY.

To maintain notational simplicity, in the rest of this book we will dispense with the tildes over linguistic variables and values. The fact that these are linguistic quantities will be assumed. Thus, Rule j in (3.2) will simply be written

$$R_j \quad \text{If } x_1 \text{ is } P_1^k \text{ and } x_2 \text{ is } P_2^l \text{ and} \cdots \text{and } x_n \text{ is } P_n^m\text{, then } y \text{ is } Q^j$$

where it will be understood that x_1 is a linguistic variable, P_1^k is the fuzzy set associated with linguistic value $\tilde{P}_1^k$, and so on.

3.2 FUZZY SYSTEMS

The fuzzy systems considered in this book have n inputs $x_i \in \mathcal{X}_i$, where $i = 1, 2, \ldots, n$ and $\mathcal{X}_i$ is the universe of discourse for x_i, and one output $y \in \mathcal{Y}$, where $\mathcal{Y}$ is the universe of discourse for y (as explained above, we assume a MISO fuzzy system). The fuzzy system has the following structure (Fig. 3.1):

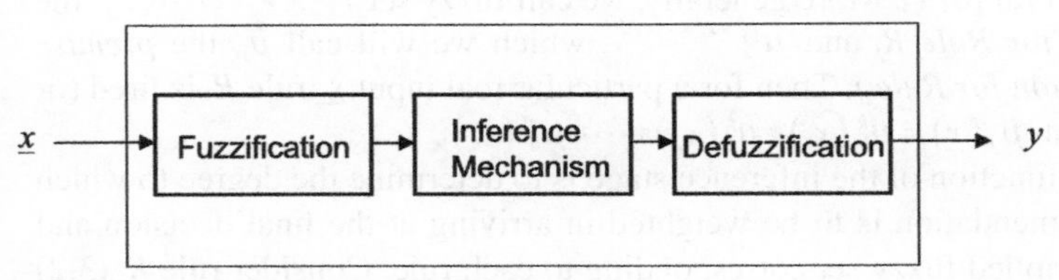

Figure 3.1. Structure of fuzzy systems.

The inputs $\underline{x}$ and output y are *crisp* (i.e., they are real numbers, not fuzzy sets). The fuzzification block converts the crisp inputs into fuzzy sets. The inference mechanism uses the rules in the rule base to convert these fuzzy sets into other fuzzy sets that are representative of the recommendations of the various rules in the rule base. The defuzzification block combines these fuzzy recommendations to give a crisp output y.

3.3 FUZZIFICATION

The function of the fuzzification stage is to convert the measured quantities from the process (voltages, velocities, temperatures, etc.) into fuzzy sets to be used by the inference stage. For example, if there is process or measurement noise, we may want to account for this by creating fuzzy sets for the measured quantities rather than assuming they are accurate as measured. In this case, the measured quantities are not believed exactly as measured (because they contain noise), but are converted into fuzzy sets that reflect their degree of undependability.

In many cases, however, the measurements are believed as measured. If measurement and process noise is low and measured quantities can be taken as true, the fuzzification stage consists of creating singleton membership functions at the measured quantities (see Section 2.5). Singleton fuzzifiucation will be used throughout the remainder of this book.

3.4 INFERENCE

The first function of the inference stage is to determine the degree of firing of each rule in the rule base. Consider rule R_i (3.1), which has a single input x. Let fuzzy set P be characterized by the membership function $\mu^P(x)$, and fuzzy set Q be characterized by the membership function $\mu^Q(y)$. For a particular crisp input $x \in \mathcal{X}$, we say rule R_i is *fired*, or is *on* (i.e., it is taken as true) to the extent $\mu^P(x)$. As mentioned in Chapter 2, this is a real number in the interval [0, 1]. More generally, we call fuzzy set P the *premise fuzzy set for Rule i*, and μ^P, which we will call μ_i, the *premise membership function for Rule i*. Then, for a particular real input x, rule R_i is fired (or is *on*) to the extent $\mu_i(x)$.

Now consider rule R_j (3.2), which has n inputs $x_1, x_2, \ldots, x_n$. Let the fuzzy set $P_1^k \times P_2^l \times \cdots \times P_n^m$ be characterized by the membership function $\mu^{P_1^k \times P_2^l \times \cdots \times P_n^m}$. For a particular real input $\underline{x} = (x_1, x_2, \ldots, x_n) \in \mathcal{X}_1 \times \mathcal{X}_2 \times \ldots \times \mathcal{X}_n$, we say rule R_j is *fired* to the extent $\mu^{P_1^k \times P_2^l \times \cdots \times P_n^m}(\underline{x}) = \mu_1^k(x_1) * \mu_2^l(x_2) * \cdots * \mu_n^m(x_n)$ [see (2.5)]. This is a real number in the interval [0, 1]. More generally, we call fuzzy set $P_1^k \times P_2^l \times \cdots \times P_n^m$ the *premise fuzzy set for Rule* R_j and $\mu^{P_1^k \times P_2^l \times \cdots \times P_n^m}$, which we will call μ_j, the *premise membership function for Rule j*. Then for a particular real input $\underline{x}$, rule R_j is fired (or is *on*) to the extent $\mu_j(\underline{x}) = \mu_1^k(x_1) * \mu_2^l(x_2) * \cdots * \mu_n^m(x_n)$.

The second function of the inference stage is to determine the degree to which each rule's recommendation is to be weighted in arriving at the final decision and to determine an implied fuzzy set corresponding to each rule. Consider rule R_j (3.2) with input $\underline{x} = (x_1, x_2, \ldots, x_n)$. From the discussion above, we know this rule is fired to the degree $\mu_j(\underline{x})$. Therefore we *attenuate* the recommendation of rule R_j, which is fuzzy set Q^j characterized by $\mu^{Q^j}(y)$, by $\mu_j(\underline{x})$. This produces an *implied fuzzy set* $\hat{Q}^j$ defined on $\mathcal{Y}$, characterized by the membership function

$$\mu^{\hat{Q}^j}(y) = \mu_j(\underline{x}) * \mu^{Q^j}(y) \tag{3.3}$$

If there are R rules in the form of (3.2), each rule has its own premise membership function $\mu_j(\underline{x}), j = 1, 2, \ldots, R$. The R rules produce R implied fuzzy sets $\hat{Q}^j, j = 1, 2, \ldots, R$, each characterized by a membership function calculated as in (3.3). Note that $\mu^{Q^j}(y)$ and $\mu^{\hat{Q}^j}(y)$ are defined $\forall y \in \mathcal{Y}$. On the other hand, the degree of firing of rule j, $\mu_j(\underline{x})$, is a function of a particular real vector $\underline{x} \in \Re^n$, hence is a real number.

In summary, the function of the inference stage is twofold: (1) to determine the degree of firing of each rule in the rule base, and (2) to create an implied fuzzy set for each rule corresponding to the rule's degree of firing.

3.5 DEFUZZIFICATION

The function of the defuzzification stage is to convert the collection of recommendations of all rules into a crisp output. Consider a rule base consisting of R rules of the form (3.2). Then, we have R implied fuzzy sets, one from each rule, each recommending a particular outcome. In order to arrive at one crisp output y, we combine all of these recommendations by taking a weighted average of the various recommendations. There are several ways to do this, but perhaps the two most common are *center of gravity* (COG) defuzzification, and *center average* (CA) defuzzification.

3.5.1 Center of Gravity (COG) Defuzzification

Suppose the consequent fuzzy set of Rule i is Q^i, characterized by membership $\mu^{Q^i}(y)$. Define the *center of area* of $\mu^{Q^i}(y)$ to be the point q_i in the universe $\mathcal{Y}$ with the property that

$$\int_{-\infty}^{q_i} \mu^{Q^i}(y)dy = \int_{q^i}^{\infty} \mu^{Q^i}(y)dy \tag{3.4}$$

Then, the crisp output of the fuzzy system is calculated using COG defuzzification as

$$y^{\text{crisp}} = \frac{\sum_{i=1}^{R} q_i \int \mu^{\hat{Q}^i}}{\sum_{i=1}^{R} \int \mu^{\hat{Q}^i}} \tag{3.5}$$

where $\int \mu^{\hat{Q}^i}$ is the area under $\mu^{\hat{Q}^i}$. The expression in (3.5) is a weighted average of the q_i's, $i = 1, \ldots, R$.

3.5.2 Center Average (CA) Defuzzification

Suppose the consequent fuzzy set of Rule i is Q^i, characterized by membership $\mu^{Q^i}(y)$. Then the crisp output of the fuzzy system is calculated using CA defuzzification as

$$y^{\text{crisp}} = \frac{\sum_{i=1}^{R} q_i \max_y \left\{\mu^{\hat{Q}^i}(y)\right\}}{\sum_{i=1}^{R} \max_y \left\{\mu^{\hat{Q}^i}(y)\right\}} \tag{3.6}$$

Note that if Q^i is normal (it usually is), then $\max_y \left\{\mu^{\hat{Q}^i}(y)\right\} = \mu_i(\underline{x})$ whether *min* or *prod* is used as the T-norm in (3.3). In that case, CA defuzzification gives

$$y^{\text{crisp}} = \frac{\sum_{i=1}^{R} q_i \mu_i(\underline{x})}{\sum_{i=1}^{R} \mu_i(\underline{x})} \tag{3.7}$$

3.6 EXAMPLE: FUZZY SYSTEM FOR WIND CHILL

Consider a fuzzy system to determine what the temperature feels like to a person under certain weather conditions. This is known as "wind chill." Wind chill has to do with the rate at which heat is removed from the person's body—the lower the wind chill, the faster heat is removed. Wind chill cannot make a body any colder than the outside temperature; it only feels colder because heat is being removed at a faster rate.

Assume that an "expert" (i.e., someone who is outside a lot) says the wind chill is determined by the actual temperature and the wind speed (it actually depends on humidity also, but we omit that for this example). Suppose he/she has identified four fuzzy sets for TEMPERATURE:COLD, COOL, WARM, and HOT (characterized in Fig. 3.2), and three fuzzy sets for WIND SPEED:LOW, GENTLE, and HIGH (characterized in Fig. 3.3).

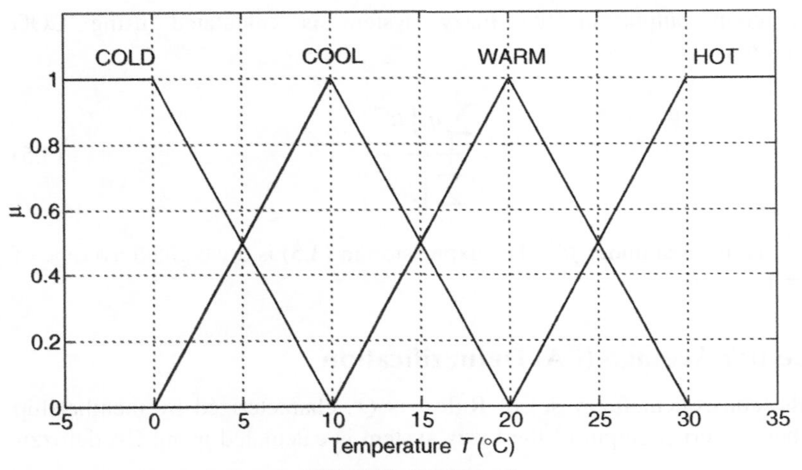

Figure 3.2. Fuzzy sets on the TEMPERATURE universe.

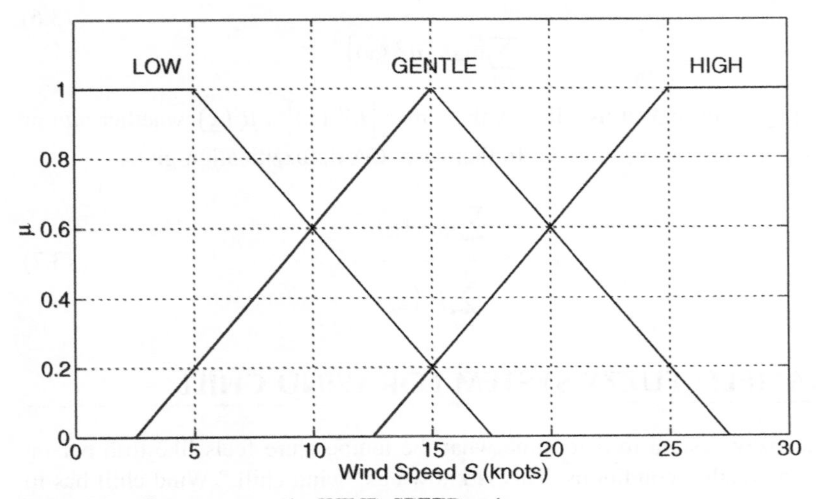

Figure 3.3. Fuzzy sets on the WIND SPEED universe.

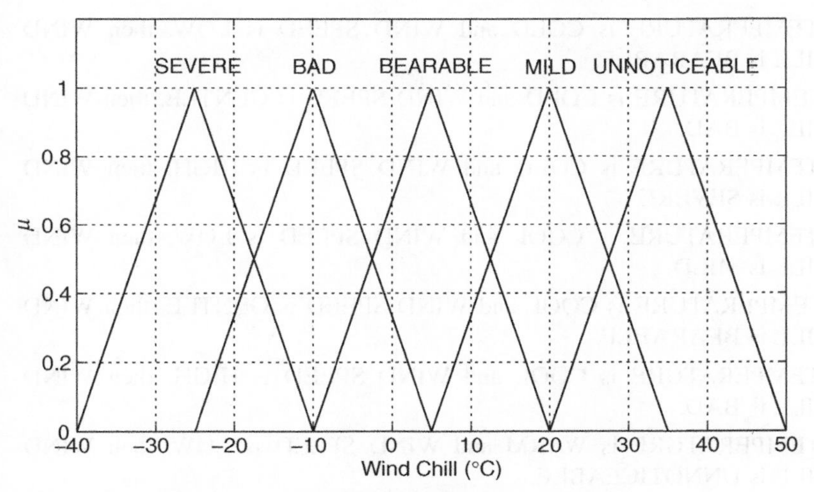

Figure 3.4. Fuzzy sets on the WIND CHILL universe.

Suppose also that he/she has identified five fuzzy sets for WIND CHILL:SEVERE, BAD, BEARABLE, MILD, and UNNOTICEABLE, characterized on the output universe of discourse as in Figure 3.4.

Notice from Figure 3.4 that the end memberships on the output universe (i.e., SEVERE and UNNOTICEABLE) do not saturate at 1 as the end memberships on the input universes do. To see why, look at the operations involved in the defuzzification step. The above defuzzification methods require the calculation of the areas of the implied fuzzy sets $\int \mu^{\hat{Q}^i}$ (for COG defuzzification) and the centers of area q_i of the output fuzzy sets (for both COG and CA). Both of these calculations are ill-defined if the output fuzzy sets saturate at 1.

Also, note that the output of the fuzzy system is a *dependent* variable, while the inputs to the fuzzy system are *independent* variables. Therefore, we must allow for the inputs to be any values, but the outputs are prescribed by the fuzzy system, hence their range is restricted. For example, while it may be unlikely, we nevertheless must allow for the *possibility* that the outdoor temperature could go significantly less than 0°C or significantly greater than 30°C. In fact, we have no control whatever over the outdoor temperature (it is an *independent* variable).

Conversely, the lowest wind chill our fuzzy system will be capable of outputting is −25°C because this is the center of area of the SEVERE output fuzzy set. The output of the fuzzy system will never get below this value no matter what the outdoor temperature and wind speed are because of the way the system has been designed. A similar statement could be made about the UNNOTICEABLE fuzzy set. Therefore our Wind Chill fuzzy system is only accurate for wind chills between −25 and 30°C. A larger range could be designed if desired, but it too will be limited to some practical range of wind chills.

Suppose finally that the "expert" has given us the following common-sense rules for determining the wind chill:

1. If TEMPERATURE is COLD and WIND SPEED is LOW, then WIND CHILL is BEARABLE.
2. If TEMPERATURE is COLD and WIND SPEED is GENTLE, then WIND CHILL is BAD.
3. If TEMPERATURE is COLD and WIND SPEED is HIGH, then WIND CHILL is SEVERE.
4. If TEMPERATURE is COOL and WIND SPEED is LOW, then WIND CHILL is MILD.
5. If TEMPERATURE is COOL and WIND SPEED is GENTLE, then WIND CHILL is BEARABLE.
6. If TEMPERATURE is COOL and WIND SPEED is HIGH, then WIND CHILL is BAD.
7. If TEMPERATURE is WARM and WIND SPEED is LOW, then WIND CHILL is UNNOTICEABLE.
8. If TEMPERATURE is WARM and WIND SPEED is GENTLE, then WIND CHILL is MILD.
9. If TEMPERATURE is WARM and WIND SPEED is HIGH, then WIND CHILL is BEARABLE.
10. If TEMPERATURE is HOT and WIND SPEED is LOW, then WIND CHILL is UNNOTICEABLE.
11. If TEMPERATURE is HOT and WIND SPEED is GENTLE, then WIND CHILL is UNNOTICEABLE.
12. If TEMPERATURE is HOT and WIND SPEED is HIGH, then WIND CHILL is MILD.

Now the fuzzy system is completely specified (i.e., we have specified all premise and consequent fuzzy sets, together with the rule base). Note that there are 12 rules, one for each combination of TEMPERATURE and WIND SPEED. This kind of rule base is said to be *complete*. For conciseness, the rule base can be tabulated as in Table 3.1:

TABLE 3.1 WIND CHILL Corresponding to Various Combinations of TEMPERATURE and WIND SPEED[a]

Wind Chill		WIND SPEED		
		LOW	GENTLE	HIGH
Temperature	COLD	BEARABLE	BAD	SEVERE
	COOL	MILD	BEARABLE	BAD
	WARM	UNNOTICEABLE	MILD	BEARABLE
	HOT	UNNOTICEABLE	UNNOTICEABLE	MILD

[a]Tabulation of above rule base.

3.6.1 Wind Chill Calculation, *Minimum* T-Norm, COG Defuzzification

Let us calculate the wind chill corresponding to a temperature of 7°C and a wind speed of 22 knots, using *min* T-norm and COG defuzzification.

Inference

To evaluate the degree to which a temperature of 7°C qualifies as COLD, COOL, WARM, and HOT, evaluate these fuzzy sets' membership functions for $T = 7$. Referring to Figure 3.2, we have

$$\mu^{\text{COLD}}(7) = 0.3$$
$$\mu^{\text{COOL}}(7) = 0.7$$
$$\mu^{\text{WARM}}(7) = 0$$
$$\mu^{\text{HOT}}(7) = 0$$

Similarly, the degree to which a wind speed of 22 knots qualifies as LOW, GENTLE, and HIGH is calculated from Figure 3.3 as

$$\mu^{\text{LOW}}(22) = 0$$
$$\mu^{\text{GENTLE}}(22) = 0.44$$
$$\mu^{\text{HIGH}}(22) = 0.76$$

Using *minimum* T-norm, we obtain the following degrees of firing for the 12 rules in the rule base:

$$R_1 : \mu_1(7,22) = \mu^{\text{COLD}\cap\text{LOW}} = \min(0.3, 0) = 0$$
$$R_2 : \mu_2(7,22) = \mu^{\text{COLD}\cap\text{GENTLE}} = \min(0.3, 0.44) = 0.3$$
$$R_3 : \mu_3(7,22) = \mu^{\text{COLD}\cap\text{HIGH}} = \min(0.3, 0.76) = 0.3$$
$$R_4 : \mu_4(7,22) = \mu^{\text{COOL}\cap\text{LOW}} = \min(0.7, 0) = 0$$
$$R_5 : \mu_5(7,22) = \mu^{\text{COOL}\cap\text{GENTLE}} = \min(0.7, 0.44) = 0.44$$
$$R_6 : \mu_6(7,22) = \mu^{\text{COOL}\cap\text{HIGH}} = \min(0.7, 0.76) = 0.7$$
$$R_7 : \mu_7(7,22) = \mu^{\text{WARM}\cap\text{LOW}} = \min(0, 0) = 0$$
$$R_8 : \mu_8(7,22) = \mu^{\text{WARM}\cap\text{GENTLE}} = \min(0, 0.44) = 0$$
$$R_9 : \mu_9(7,22) = \mu^{\text{WARM}\cap\text{HIGH}} = \min(0, 0.76) = 0$$
$$R_{10} : \mu_{10}(7,22) = \mu^{\text{HOT}\cap\text{LOW}} = \min(0, 0) = 0$$
$$R_{11} : \mu_{11}(7,22) = \mu^{\text{HOT}\cap\text{GENTLE}} = \min(0, 0.44) = 0$$
$$R_{12} : \mu_{12}(7,22) = \mu^{\text{HOT}\cap\text{HIGH}} = \min(0, 0.76) = 0$$

Thus we see that for this input $(T, S) = (7, 22)$, Rule 6 is fired with the greatest certainty, Rule 5 is fired with less certainty, and so on. Many rules are not fired at all.

The membership arrangement in Figure 3.2 is called a *partition of unity* because the sum of all memberships equals 1 at every T. The membership arrangement in Figure 3.3 is not a partition of unity. Note that a group of Gaussian membership functions can never form a partition of unity.

Partitions of unity simplify defuzzification (see Chapter 4), but it is not absolutely necessary to use them. As always, fuzzy sets should be defined so that they most accurately reflect the quantities they describe, not merely to simplify calculations.

Since only four rules are fired for this input, we will create four nonzero implied fuzzy sets. To create the implied fuzzy sets for Rules 2, 3, 5, and 6, we attenuate each rule's consequent membership function by the degree of firing of the rule. Using *minimum* T-norm, we obtain the membership function characterizing Rule 2's implied fuzzy set as

$$\mu_2^{\text{implied}}(y) = \min_y \{\mu_2(7,22), \mu^{\text{BAD}}(y)\}$$

This minimum is taken pointwise in $\mathcal{Y}$. Similarly, the membership function characterizing Rules 3, 5, and 6's implied fuzzy sets are

$$\mu_3^{\text{implied}}(y) = \min_y \{\mu_3(7,22), \mu^{\text{SEVERE}}(y)\}$$

$$\mu_5^{\text{implied}}(y) = \min_y \{\mu_4(7,22), \mu^{\text{BEARABLE}}(y)\}$$

$$\mu_6^{\text{implied}}(y) = \min_y \{\mu_6(7,22), \mu^{\text{BAD}}(y)\}$$

The resulting implied fuzzy sets' membership functions are shown in Figure 3.5. Some of the memberships in Figure 3.5 have been slightly displaced so they can be seen more clearly.

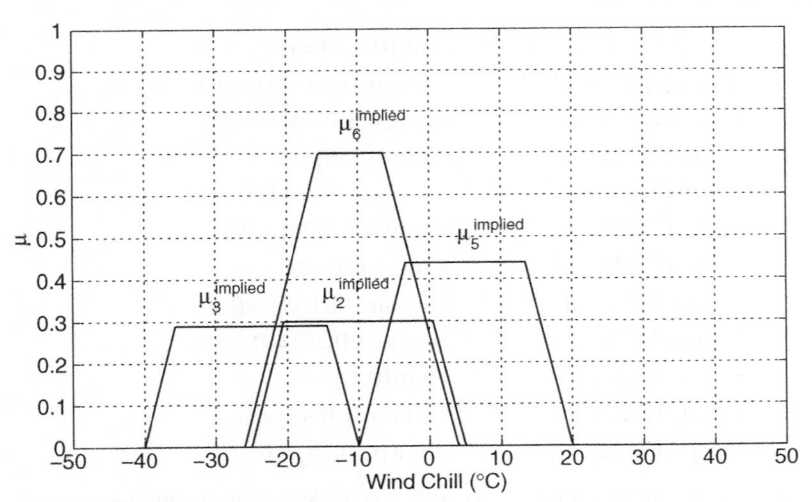

Figure 3.5. Implied fuzzy sets, *minimum* T-norm (slightly displaced to improve clarity).

Defuzzification

To calculate a crisp output using COG defuzzification, it is necessary to find the centers of area of the memberships characterizing the output fuzzy sets [q_i, $i = 2, 3, 5$, and 6 in (3.5)] and the areas of the four trapezoidal membership functions characterizing the implied fuzzy sets in Figure 3.5 [$\int \mu_i^{\text{implied}}$, $i = 2, 3, 5, 6$ in (3.5)]. These are

$$q_2 = -10$$

$$q_3 = -25$$

$$q_5 = 5$$

$$q_6 = -10$$

$$\int \mu_2^{\text{implied}} = 7.65$$

$$\int \mu_3^{\text{implied}} = 7.65$$

$$\int \mu_5^{\text{implied}} = 10.296$$

$$\int \mu_6^{\text{implied}} = 13.65$$

Note: The area of a trapezoid with base w and height h is $w(h - h^2/2)$.

Therefore, the crisp output of the fuzzy system corresponding to the crisp input $(T, S) = (7, 22)$, is calculated using COG defuzzification to be

$$y^{\text{crisp}} = \frac{-10(7.65) - 25(7.65) + 5(10.296) - 10(13.65)}{7.65 + 7.65 + 10.296 + 13.65} = -8.9887°\text{C} \quad (3.8)$$

Thus, the wind chill corresponding to a temperature of 7°C and a wind speed of 22 knots, calculated using the rule base and fuzzy sets specified above, using the *minimum* T-norm for conjunction in the premise and inference, and COG defuzzification, is –8.9887°C.

Since there are only two inputs, the input–output characteristic of the fuzzy system can be plotted. The characteristic is shown in Figure 3.6.

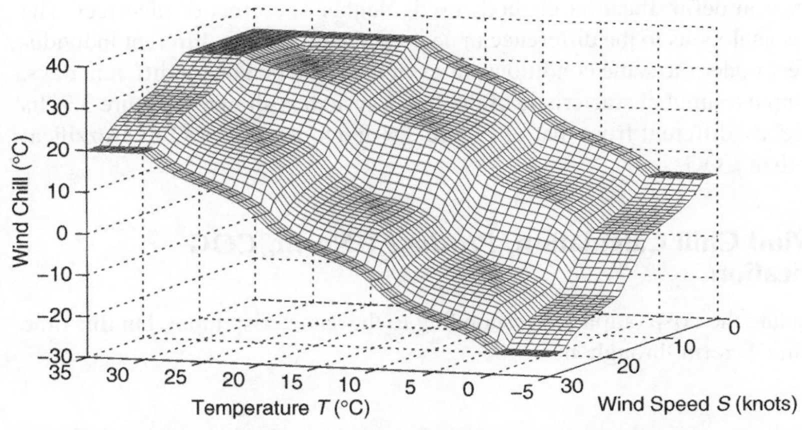

Figure 3.6. Input–output characteristic of *Wind Chill* fuzzy system, *min* T-norm, COG defuzzification.

The *general* shape of the characteristic reflects our common sense about wind chill: that is that higher wind chill temperatures result from higher temperatures combined with lower wind speeds, and lower wind chill temperatures result from lower temperatures combined with higher wind speeds. The *local* shape of the characteristic, that is, the waves and undulations, is a function of the membership shapes, T-norm, and defuzzification method used. Therefore, the local shape is not very important; the global shape is.

The flat areas at the extremes of the characteristic reflect the fact that the leftmost and rightmost TEMPERATURE memberships saturate at 0 and 30°C, respectively, the leftmost and rightmost WIND SPEED memberships saturate at 2.5 and 27.5 kn, respectively, and the leftmost and rightmost WIND CHILL membership centers are at −25 and 35°C, respectively. Therefore, the characteristic is only accurate for (T, S) such that $0°\text{C} \leq T \leq 30°\text{C}$ and $2.5 \leq S \leq 27.5$ kn. If the characteristic is examined closely, it can be seen that the point $(T, S) = (7, 22)$ corresponds to a wind chill of −8.9887°C.

3.6.2 Wind Chill Calculation, *Minimum* T-Norm, CA Defuzzification

If CA defuzzification is used instead of COG, the crisp output of the fuzzy system corresponding to the crisp input $(T, S) = (7, 22)$ is calculated to be

$$y^{\text{crisp}} = \frac{-10(0.3)-25(0.3)+5(0.44)-10(0.7)}{0.3+0.3+0.44+0.7} = -8.7931°\text{C} \qquad (3.9)$$

Thus, the wind chill corresponding to a temperature of 7°C and a wind speed of 22 knots, calculated using the rule base and fuzzy sets specified above, using the *minimum* T-norm for conjunction in the premise and inference, and CA defuzzification, is −8.7931°C.

The wind chill −8.7931°C calculated in (3.9) is close to the wind chill −8.9887°C calculated in (3.8), but not exactly equal to it. The difference is due to the difference in defuzzification methods used. Neither is correct or incorrect. The difference is analogous to the difference in percieved temperature different individuals might feel under the same conditions, or even the same person on different days.

The input–output characteristic of the fuzzy system is shown in Figure 3.7. Its shape is slightly different from that of Figure 3.6 due to the use of CA defuzzification rather than COG.

3.6.3 Wind Chill Calculation, *Product* T-Norm, COG Defuzzification

Let us calculate the crisp output corresponding to the same crisp input, but this time using *product* T-norm throughout.

Inference

Since we are using the same input as before $(T, S) = (7, 22)$, we have the same degrees of belongingness for all input fuzzy sets as before. Using *product* T-norm, we obtain the following degrees of firing for the rules:

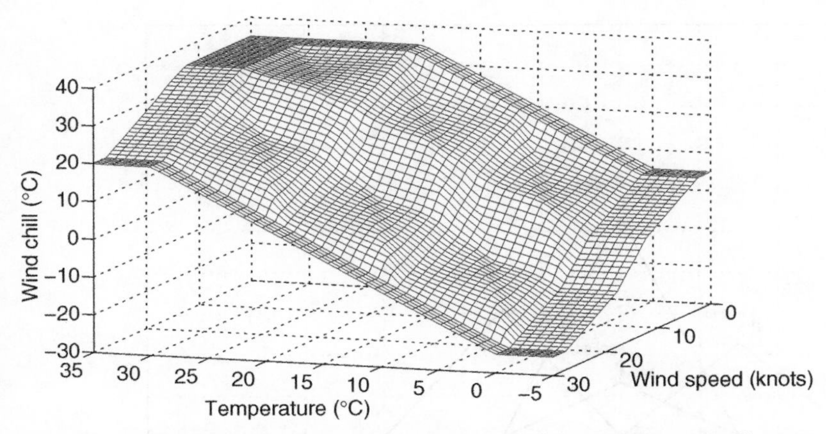

Figure 3.7. Input–output characteristic of *Wind Chill* fuzzy system, *min* T-norm, CA defuzzification.

$$R_1 : \mu_1(7,22) = \mu^{\mathrm{COLD} \cap \mathrm{LOW}} = (0.3)(0) = 0$$
$$R_2 : \mu_2(7,22) = \mu^{\mathrm{COLD} \cap \mathrm{GENTLE}} = (0.3)(0.44) = 0.132$$
$$R_3 : \mu_3(7,22) = \mu^{\mathrm{COLD} \cap \mathrm{HIGH}} = (0.3)(0.76) = 0.228$$
$$R_4 : \mu_4(7,22) = \mu^{\mathrm{COOL} \cap \mathrm{LOW}} = (0.7)(0) = 0$$
$$R_5 : \mu_5(7,22) = \mu^{\mathrm{COOL} \cap \mathrm{GENTLE}} = (0.7)(0.44) = 0.308$$
$$R_6 : \mu_6(7,22) = \mu^{\mathrm{COOL} \cap \mathrm{HIGH}} = (0.7)(0.76) = 0.532$$
$$R_7 : \mu_7(7,22) = \mu^{\mathrm{WARM} \cap \mathrm{LOW}} = (0)(0) = 0$$
$$R_8 : \mu_8(7,22) = \mu^{\mathrm{WARM} \cap \mathrm{GENTLE}} = (0)(0.44) = 0$$
$$R_9 : \mu_9(7,22) = \mu^{\mathrm{WARM} \cap \mathrm{HIGH}} = (0)(0.76) = 0$$
$$R_{10} : \mu_{10}(7,22) = \mu^{\mathrm{HOT} \cap \mathrm{LOW}} = (0)(0) = 0$$
$$R_{11} : \mu_{11}(7,22) = \mu^{\mathrm{HOT} \cap \mathrm{GENTLE}} = (0)(0.44) = 0$$
$$R_{12} : \mu_{12}(7,22) = \mu^{\mathrm{HOT} \cap \mathrm{HIGH}} = (0)(0.76) = 0$$

Again, only Rules 2, 3, 5, and 6 are fired. Using *product* T-norm, we obtain the membership functions characterizing Rules 2, 3, 5, and 6's implied fuzzy sets as

$$\mu_2^{\mathrm{implied}}(y) = (\mu_2(7,22))(\mu^{\mathrm{BAD}}(y))$$
$$\mu_3^{\mathrm{implied}}(y) = (\mu_3(7,22))(\mu^{\mathrm{SEVERE}}(y))$$
$$\mu_5^{\mathrm{implied}}(y) = (\mu_5(7,22))(\mu^{\mathrm{BEARABLE}}(y))$$
$$\mu_6^{\mathrm{implied}}(y) = (\mu_6(7,22))(\mu^{\mathrm{BAD}}(y))$$

The resulting membership functions are shown in Figure 3.8.

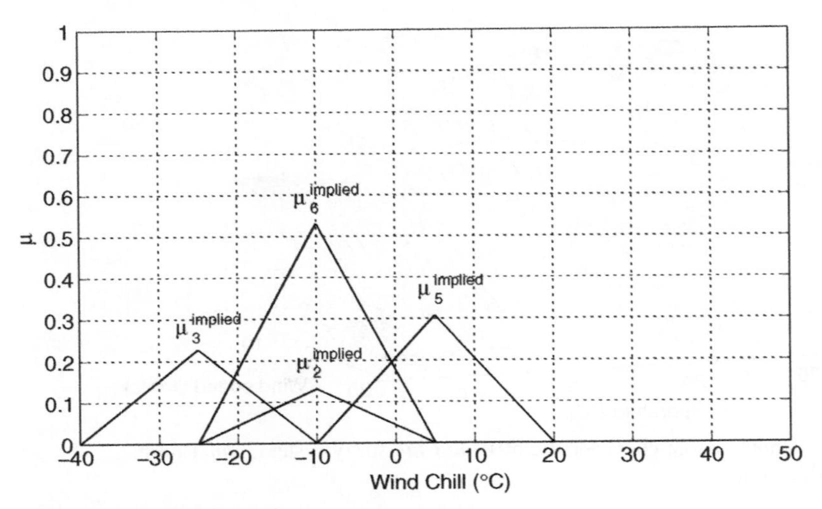

Figure 3.8. Implied fuzzy sets, *product* T-norm.

Defuzzification

The centers of area of the output fuzzy sets [q_i, $i = 2, 3\ 5, 6$ in (3.5)] are the same as before, and the areas of the four triangular membership functions characterizing the implied fuzzy sets in Figure 3.8 [$\int \mu_i^{\text{implied}}$, $i = 2, 3\ 5$, and 6 in (3.5)] are

$$\int \mu_2^{\text{implied}} = 1.98$$

$$\int \mu_3^{\text{implied}} = 3.42$$

$$\int \mu_5^{\text{implied}} = 4.62$$

$$\int \mu_6^{\text{implied}} = 7.98$$

Therefore, the crisp output of the fuzzy system corresponding to the crisp input $(T, S) = (7, 22)$ is calculated using COG defuzzification to be

$$y^{\text{crisp}} = \frac{-10(1.98) - 25(3.42) + 5(4.62) - 10(7.98)}{1.98 + 3.42 + 4.62 + 7.98} = -9.0°\text{C} \tag{3.10}$$

Thus, the wind chill corresponding to a temperature of 7°C and a wind speed of 22 knots, calculated using the rule base and fuzzy sets specified above, using the *product* T-norm for conjunction in the premise and inference, and COG defuzzification, is −9.0°C. The input–output characteristic of the fuzzy system is shown in Figure 3.9. This is slightly different from the previous characteristics (Figs. 3.6 and 3.7) due to the different T-norm and defuzzification methods. The characteristic still has the same general shape as the others.

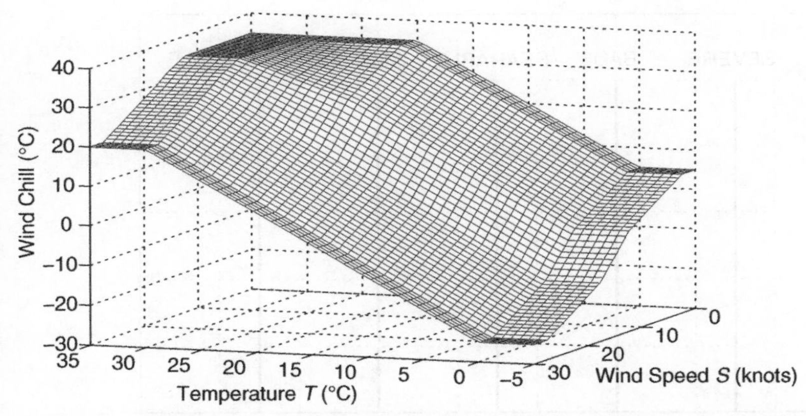

Figure 3.9. Input–output characteristic of *Wind Chill* fuzzy system, *product* T-norm, COG defuzzification.

3.6.4 Wind Chill Calculation, *Product* T-Norm, CA Defuzzification

If CA defuzzification is used instead of COG, the crisp output of the fuzzy system corresponding to the crisp input $(T, S) = (7, 22)$ is calculated to be

$$y^{\text{crisp}} = \frac{-10(0.132) - 25(0.228) + 5(0.308) - 10(0.532)}{0.132 + 0.228 + 0.308 + 0.532} = -9.0°\text{C} \quad (3.11)$$

Thus, the wind chill corresponding to a temperature of 7°C and a wind speed of 22 knots, calculated using the rule base and fuzzy sets specified above, using the *product* T-norm for conjunction in the premise and inference, and CA defuzzification, is −9.0°C. The input–output characteristic is identical to that of Figure 3.9. This is due to the fact that the areas of the triangles in Figure 3.8 are proportional to the premise membership values of the corresponding rules. This is not true for the trapezoidal memberships of Figure 3.5.

3.6.5 Wind Chill Calculation, Singleton Output Fuzzy Sets, *Product* T-Norm, CA Defuzzification

Another possibility for output fuzzy sets is *singleton* fuzzy sets. As mentioned in Section 2.5, these are characterized by membership functions that are zero except for one point in the output universe, where they are 1. For example, the output fuzzy sets SEVERE, BAD, BEARABLE, MILD, and UNNOTICEABLE could be characterized by the membership functions shown in Figure 3.10.

Now, defuzzification must be CA since there is no area to compute. Also, since the singletons are normal, Eq. (3.7) is always used for defuzzification. Therefore, the above results for triangular output memberships, *product* T-norm, and CA defuzzification would also be obtained if the output fuzzy sets were singletons. In general, if *product* T-norm is used and the output memberships are symmetrical and

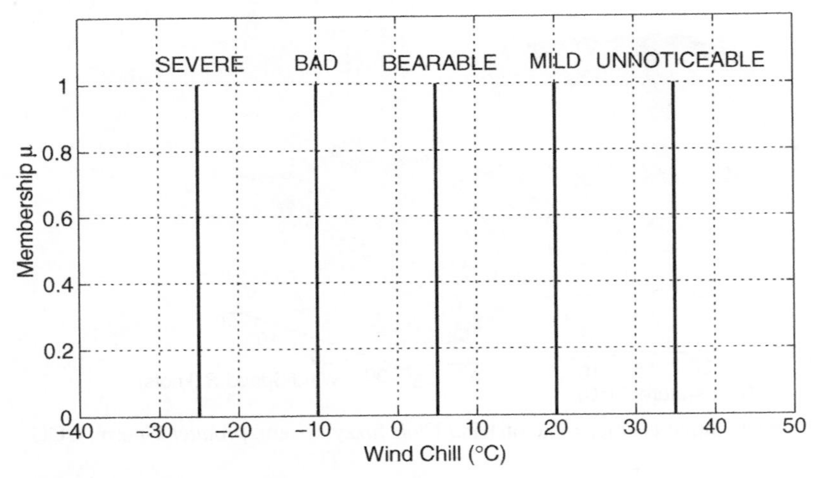

Figure 3.10. Membership functions characterizing singleton output fuzzy sets.

normal, the output fuzzy sets may be replaced by singletons. In the remaining chapters of this book, only singleton output fuzzy sets will be used.

3.7 SUMMARY

A fuzzy system uses fuzzy reasoning processes to convert crisp inputs into crisp outputs. The main components of the fuzzy system are a **fuzzification** section, an **inference** mechanism, and a **defuzzification** section. A set of rules generally in **if-then** *modus ponens* form, called a **rule base**, specifies how decisions are to be made based on the measured inputs. In Mamdani fuzzy systems, the consequent of each rule is a fuzzy set. This is in contrast to Takagi–Sugeno (T–S) fuzzy systems (see Chapter 6), whose consequents are mathematical expressions.

The fuzzification section converts crisp inputs into fuzzy sets. There are several strategies for fuzzification, but the one most commonly used, and the only strategy used in this book, is **singleton fuzzification**. This assumes the measured input is the true input (i.e., there is no measurement noise or any other uncertainty, such as sensor bias). Thus in fuzzification the input fuzzy sets are evaluated exactly at the measured inputs.

The inference mechanism determines the extent to which each rule in the rule base applies in the present situation, and forms a corresponding implied fuzzy set for each rule. If the rule premises contain a **conjunction of several inputs**, the degree of firing of each rule is calculated by taking a T-norm of the individual members of the conjunction. The inference mechanism also calculates an **implied fuzzy set** for each rule. The membership for this set is calculated by taking a T-norm between the rule's degree of firing and the rule's output membership function.

The defuzzification section combines the implied fuzzy sets of all rules to get a crisp output. There are several ways to do this, but the only two discussed in this book are **Center of Gravity (COG)** defuzzification and **Center Average (CA)** defuzzification. Also discussed in this chapter are the **input–output characteristic** of the fuzzy system, and **singleton fuzzy sets**. Singleton fuzzy sets are often used on the output universe of discourse to simplify calculations. Singleton output fuzzy sets obviate the need for COG defuzzification, which usually gives comparable results and is much more computationally expensive to implement.

EXERCISES

3.1 Plot the input–output characteristics of Sections 3.6.1–3.6.5.

3.2 Use the *Wind Chill* fuzzy system of Section 3.6 to calculate the wind chill corresponding to a temperature of 3°C and a wind speed of 16 knots. Use *minimum* T-norm and center of gravity defuzzification. *Note*: For this wind speed, three fuzzy sets on the S universe are nonzero. Therefore there will be more than four rules "on."

3.3 Repeat Problem 3.2 using *product* T-norm and center average defuzzification.

3.4 Repeat the *Wind Chill* problem of Section 3.6 with $(T, S) = (3, 16)$ using the Gaussian fuzzy sets of Figure 2.5 for TEMPERATURE and the Gaussian fuzzy sets of Figure 2.11 for WIND SPEED. Use the triangular fuzzy sets of Figure 3.4 for the output. Use the same T-norm and defuzzification as Problem 3.2.

3.5 In Section 3.6.1, it is stated the group of Gaussian membership functions can never form a partition of unity. Specify another shape for membership functions that is similar to triangular memberships (i.e., symmetrical and normal, and corresponding to convex fuzzy sets), but has no straight lines, yet forms a partition of unity.

3.6 Consider a fuzzy system for braking a car. Let the inputs be Speed (with fuzzy sets Slow, Medium, Fast, and Very Fast) and Road Grade (with fuzzy sets Uphill, Level, and Downhill). Let the output be Brake Pressure (with fuzzy sets Zero, Light, Medium, Heavy, and Very Heavy). Specify triangular memberships forming partitions of unity for all the above linguistic values.

3.7 Give a complete rule base for a fuzzy system for braking a car. Use the fuzzy sets you defined in Problem 3.6.

3.8 In the *Wind Chill* fuzzy system of Section 3.6, give all possibilities for the number of rules that can be *on* at any one time.

3.9 A more complete fuzzy system for *Wind Chill* also includes Relative Humidity as an input in addition to Temperature and Wind Speed. Higher relative humidity causes higher wind chill. Let the input and output universes be specified as in Figs. 3.11–3.14). (a) Specify a rule base for the three-input *Wind Chill* fuzzy system. (b) Use the rule base of (a) and *product* T-norm to calculate the wind chill for the conditions (i) $(T, S, H) = (20°C, 12.5\,kn, 60\%)$, (ii) $(T, S, H) = (10°C, 18\,kn, 30\%)$, (iii) $(T, S, H) = (0°C, 25\,kn, 90\%)$.

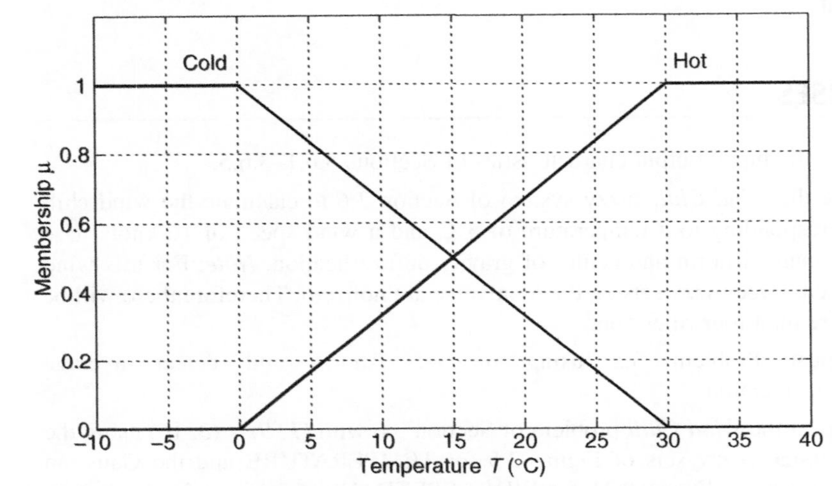

Figure 3.11. Input membership functions on Temperature universe for Problem 3.9.

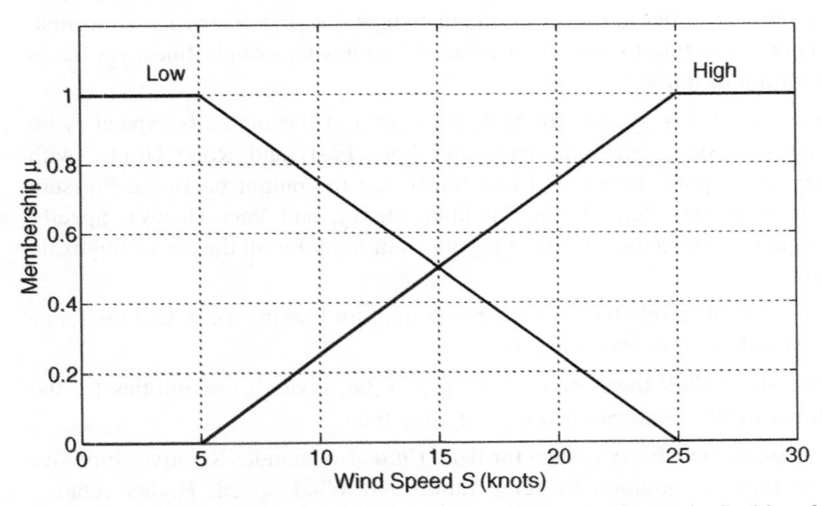

Figure 3.12. Input membership functions on Wind Speed universe for Problem 3.9.

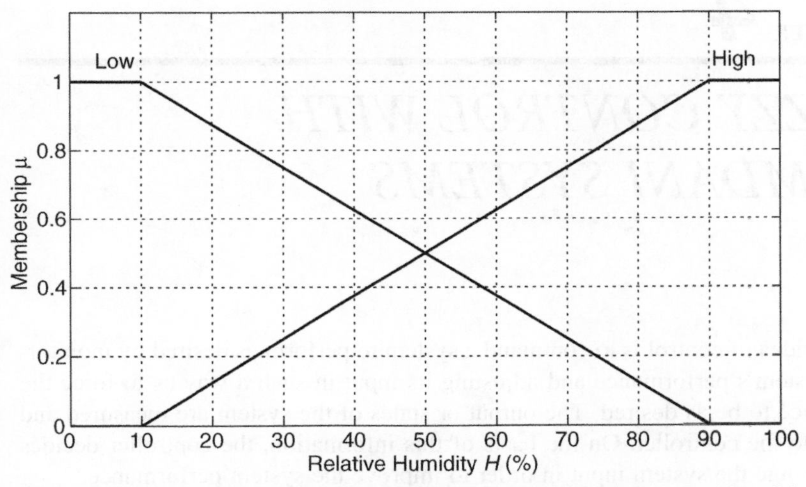

Figure 3.13. Input membership functions on Humidity universe for Problem 3.9.

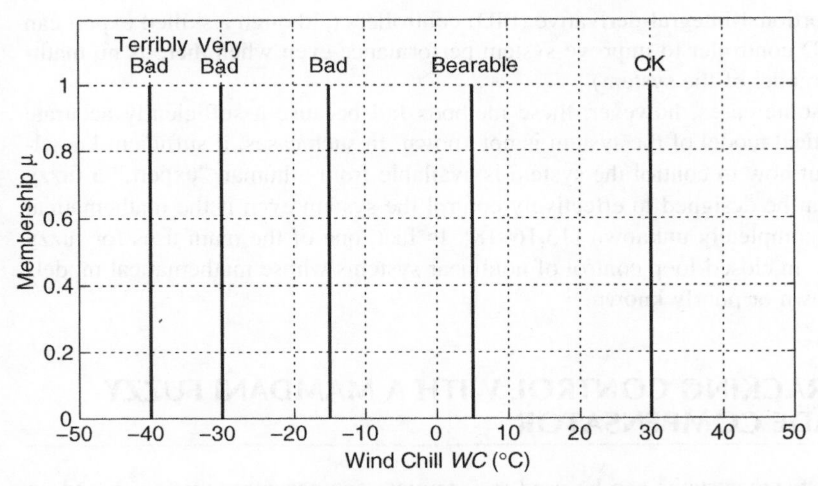

Figure 3.14. Output membership functions on Wind Chill universe for Problem 3.9.

3.10 Prove that an n-input, m-output fuzzy system is equivalent to m n-input, single-output fuzzy systems.

CHAPTER 4

FUZZY CONTROL WITH MAMDANI SYSTEMS

The basic idea of control is to command a system to perform as desired by monitoring the system's performance and adjusting its input in such a way as to force the performance to be as desired. The output or states of the system are measured and fed back to the controller. On the basis of this information, the controller decides how to change the system input in order to improve the system performance.

Much of conventional control is "model based," which means the controller design is based on a mathematical model of the system. Examples of model-based controllers are linear state feedback controllers, optimal controllers, H_∞ controllers, and proportional-integral-derivative (PID) controllers (although a skilled expert can tune a PID controller to improve system performance even when there is no mathematical model of the system).

In some cases, however, these methods fail because a sufficiently accurate mathematical model of the system is not known. In such cases, if sufficient knowledge about how to control the system is available from a human "expert," a fuzzy system can be designed to effectively control the system even if the mathematical model is completely unknown [13,16–18]. In fact, one of the main uses for fuzzy systems is in closed-loop control of nonlinear systems whose mathematical models are unknown or poorly known.

4.1 TRACKING CONTROL WITH A MAMDANI FUZZY CASCADE COMPENSATOR

Mamdani fuzzy systems can be used to formulate compensators that are based on the user's common sense about how to control a system. This kind of controller needs no mathematical model of the system, hence it is not model based. Its design is rather based on expert knowledge.

Most plants to be controlled are continuous-time, therefore their inputs and outputs are piecewise-continuous functions of time. If the control objective is tracking, the controller configuration usually involves unity feedback with a cascade compensator, as in Figure 1.6. When the compensator is a fuzzy system, the configuration of Figure 4.1 results.

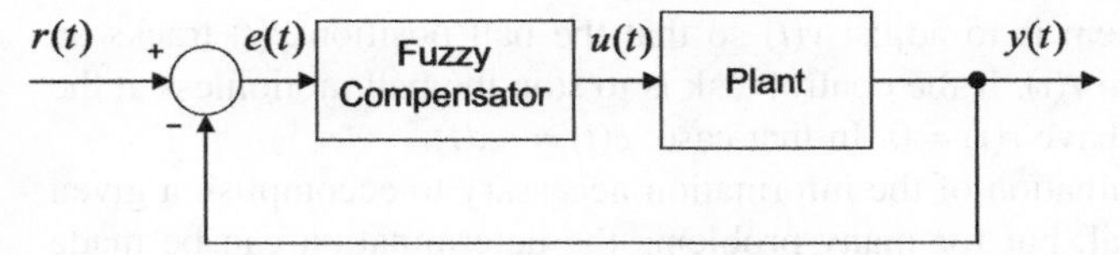

Figure 4.1. Closed-loop system with cascade fuzzy controller.

In Figure 4.1, the output of the summer is the continuous-time function $e(t) = r(t) - y(t)$, which is the tracking error. The fuzzy compensator takes $e(t)$ as its input and formulates a plant input to force the plant output to follow the reference input $r(t)$. In this chapter, the compensator will be designed from only expert knowledge about how to control the system. Therefore, no plant models will be needed.

Because the compensator is a fuzzy system, it is virtually always implemented with a digital computer. Therefore, time must be discretized with an appropriate sampling time depending on the speed of the analog signals. The input to the fuzzy compensator is a sampled version of $e(t)$, resulting in an output of the fuzzy compensator occurring at every sample time. The compensator's output must then be converted to a continuous-time signal to be fed to the plant. Because the compensator's input changes at every time step, the compensator's output also changes at every time step according to the fuzzy compensator's input–output characteristic (see Section 3.6). This characteristic does not change with time, therefore the fuzzy compensator implements a time-invariant mapping from $e(t)$ to $u(t)$ with $e(t)$ [hence $u(t)$] changing at every time step.

4.1.1 Initial Fuzzy Compensator Design: Ball and Beam Plant

Consider the ball and beam plant of Section 1.4. The system is depicted in Figure 4.2 and modeled (for simulation purposes only) by [19]:

$$\ddot{x} = 9.81 \sin kv \tag{4.1}$$

Note: The model (4.1) is actually a simplified model of the ball and beam [12].

In Figure 4.2, $x(t)$ is the position of the ball along the beam (with $x = 0$ defined as the center of the beam), and $\psi(t)$ is the beam angle commanded by the motor (with $\psi = 0$ defined as horizontal). The input to the ball and beam system is the voltage v supplied to the motor. The beam angle ψ is proportional to v (i.e., $\psi = kv$).

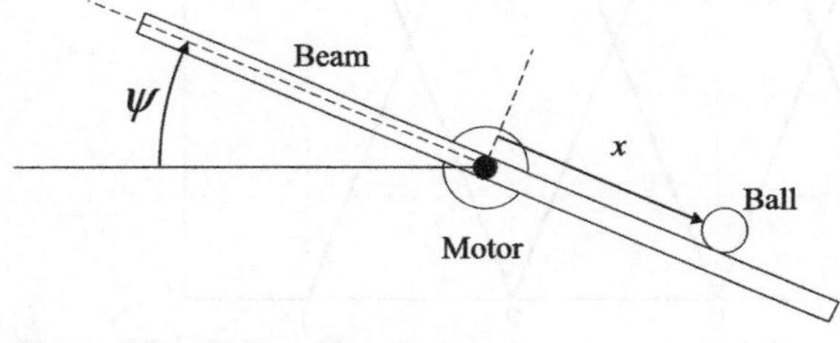

Figure 4.2. Ball and beam system.

The control problem is to adjust $v(t)$ so that the ball position $x(t)$ tracks an arbitrary reference signal $r(t)$. If the control task is to stop the ball motionless at the center of the beam, we have $r(t) = 0$. In that case, $e(t) = -x(t)$.

In general, determination of the information necessary to accomplish a given control task is not trivial, but for many problems the determination can be made using common sense. Suppose an "expert" has decided that the contol objective can be accomplished with knowledge of the ball's position and velocity. Accordingly, let the inputs to the fuzzy controller be e and $\dot{e}$, and the output be v.

Define five fuzzy sets on the e and $\dot{e}$ universes, with linguistic values *Negative Large* (NL), *Negative Small* (NS), *Zero* (Z), *Positive Small* (PS), and *Positive Large* (PL). These are characterized by the triangular memberships shown in Figures 4.3 and 4.4.

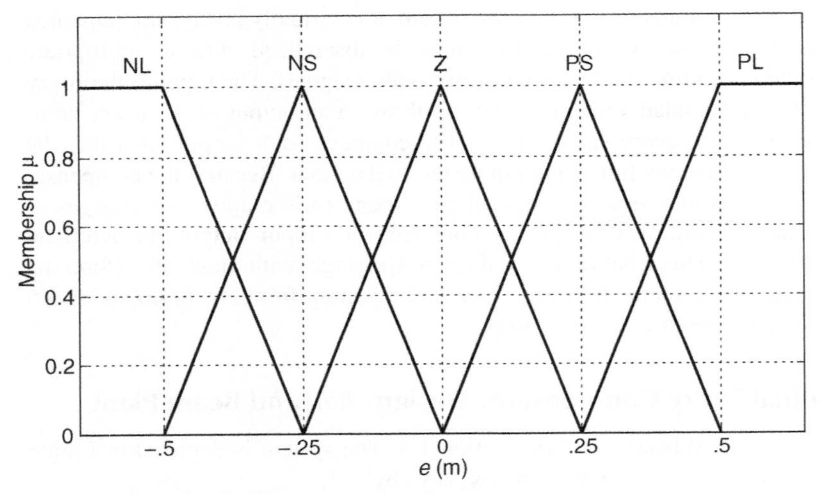

Figure 4.3. Triangular fuzzy sets on e universe.

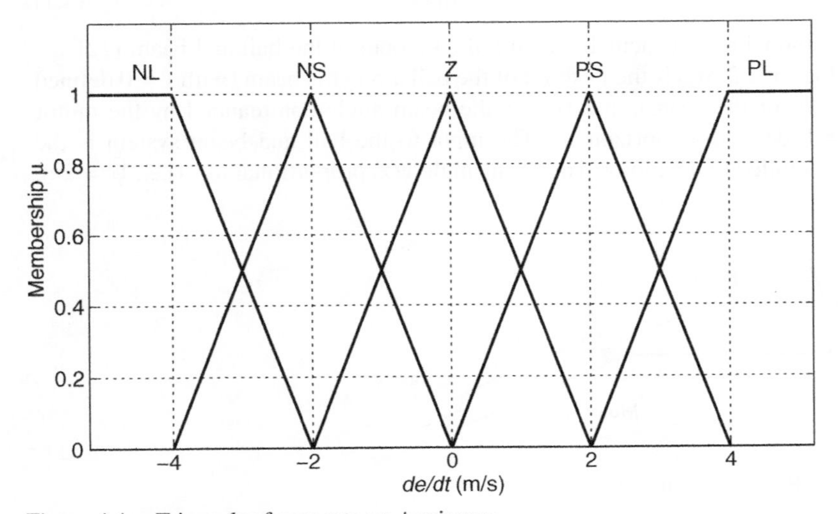

Figure 4.4. Triangular fuzzy sets on $\dot{e}$ universe.

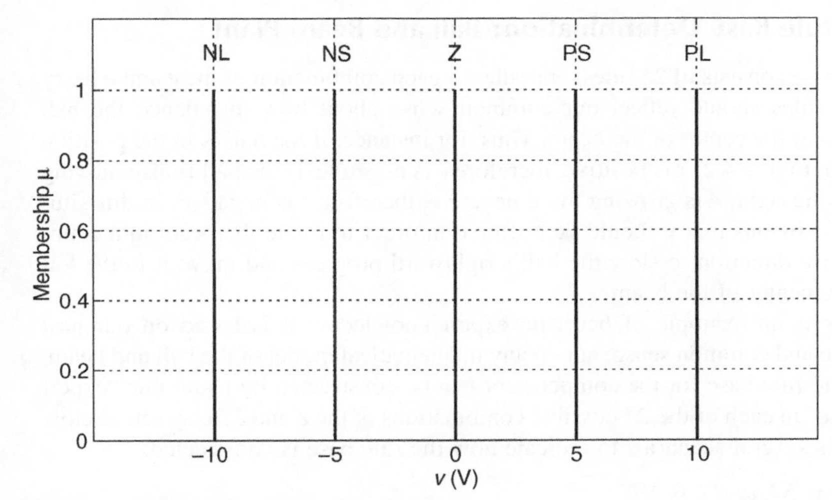

Figure 4.5. Fuzzy sets on v universe.

Similarly define five singleton fuzzy sets on the v universe, with linguistic values NL, NS, Z, PS, and PL. These are characterized by the memberships shown in Figure 4.5.

A good choice of input and output fuzzy sets is crucial to the success of any fuzzy controller. The locations of the fuzzy sets in Figure 4.3 were chosen because the beam is 1 m long with $x = 0$ at the center, therefore x (hence e) is always such that $-0.5 \leq e \leq 0.5$ m. The fuzzy sets in Figure 4.4 were chosen based on our estimation of the maximum ball velocity expected in normal operation of the beam. The effective universe of discourse $-4 \leq \dot{e} \leq 4$ m/s was arrived at by realizing that if the ball is dropped from a stationary position, its velocity will reach ~4.4 m/s when it has fallen 1 m. This is an approximation of the maximum ball velocity if it is located motionless at one end of the beam with a beam angle of $\psi = \pi/2$ (i.e., it falls vertically 1 m). This is admittedly a crude estimation, but it is important to try to find a meaningful range for $\dot{e}$, as it is for all inputs and outputs. The singleton fuzzy sets in Figure 4.5 were chosen based on our estimate of the voltage range needed to actuate the beam in order to accomplish the control task (i.e., $-10 \leq v \leq 10$V).

Recall that in Chapter 3 we saw that using singleton output fuzzy sets gave comparable results to triangular, Gaussian, or other types of output fuzzy sets under certain conditions. For some applications, especially classification [20], it may be necessary to be very meticulous about shaping output membership functions. However, for control applications, it is usually sufficient to use singleton output fuzzy sets because the controller can be otherwise adjusted by adjusting scaling gains or other parameters.

Using singleton output memberships simplifies defuzzification as well, since no areas under memberships of implied fuzzy sets need to be calculated. Efficiency of calculation is crucial for control with a digital computer due to short sampling times available for the fuzzy controller to perform its calculations. For these reasons, we will use only *product* T-norm and singleton output fuzzy sets in the remainder of this book.

4.1.2 Rule Base Determination: Ball and Beam Plant

The rule base consists of 25 rules, one rule for each combination of the e and $\dot{e}$ fuzzy sets. The rules should reflect our common sense about how to balance the ball motionless at the center of the beam. Thus, for instance, if the ball is in the position depicted in Figure 4.2, x is positive, therefore e is negative. If the ball is also moving further to the right, e is growing more negative, therefore $\dot{e}$ is negative. In this situation, it is obvious that v should be negative in order to rotate the beam in a counterclockwise direction to slow the ball's rightward progress and move it to the left toward the center of the beam.

This is an example of heuristic expert knowledge; it is based on our past experience and common sense, not on any mathematical model of the ball and beam. A complete rule base for the compensator can be constructed by using this "expert knowledge" in each of the 25 possible combinations of the e and $\dot{e}$ fuzzy sets. Below we address several scenarios to indicate how the rule base is constructed.

Case 1:* e *is NL and* ė *is NL

This situation is depicted in Figure 4.6 (the large arrow indicates fast movement to the right).

In this case, because the ball is moving quickly to the right away from the center of the beam, v should be large and negative (i.e., NL) in order to quickly rotate the beam counterclockwise to move the ball to the left.

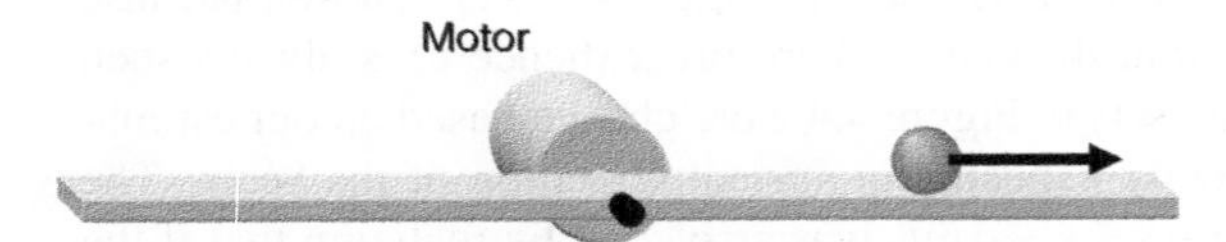

Figure 4.6. The parameter e is NL and $\dot{e}$ is NL.

Case 2:* e *is NL and* ė *is PL

This situation is depicted in Figure 4.7.

In this case, because the ball is moving quickly to the left toward the center of the beam, no rotation of the beam is necessary. Therefore, v should be zero (i.e., Z).

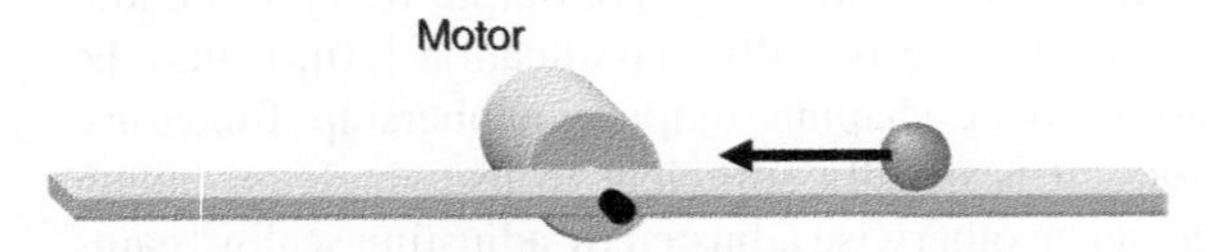

Figure 4.7. The parameter e is NL and c is PL.

Case 3:* e *is Z and* ė *is NS

This situation is depicted in Figure 4.8 (the small arrow indicates slow movement to the right).

In this case, because the ball is moving slowly to the right away from the center of the beam, a gentle counterclockwise rotation of the beam is necessary to move it leftward. Therefore, v should be a small negative amount (i.e., NS).

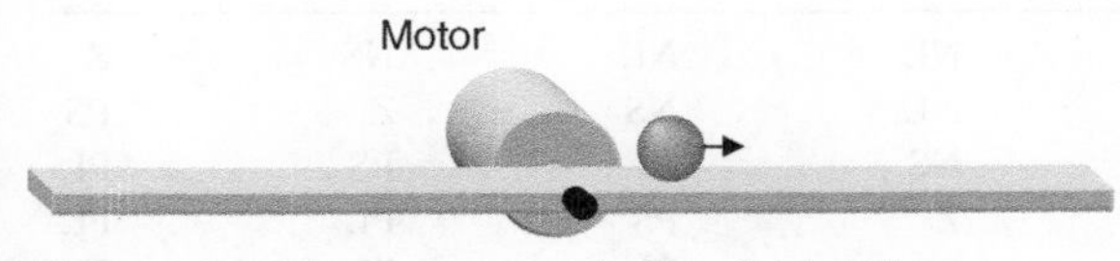

Figure 4.8. The parameter e is Z and $\dot{e}$ is NS.

The above three scenarios reflect our common sense (i.e., our "expert knowledge") about how to balance the ball motionless at the center of the beam. In this way, we can write all 25 rules in the rule base:

1. If e is NL and $\dot{e}$ is NL, then v is NL.
2. If e is NL and $\dot{e}$ is NS, then v is NL.
3. If e is NL and $\dot{e}$ is Z, then v is NL.
4. If e is NL and $\dot{e}$ is PS, then v is NS.
5. If e is NL and $\dot{e}$ is PL, then v is Z.
6. If e is NS and $\dot{e}$ is NL, then v is NL.
7. If e is NS and $\dot{e}$ is NS, then v is NL.
8. If e is NS and $\dot{e}$ is Z, then v is NS.
9. If e is NS and $\dot{e}$ is PS, then v is Z.
10. If e is NS and $\dot{e}$ is PL, then v is PS.
11. If e is Z and $\dot{e}$ is NL, then v is NL.
12. If e is Z and $\dot{e}$ is NS, then v is NS.
13. If e is Z and $\dot{e}$ is Z, then v is Z.
14. If e is Z and $\dot{e}$ is PS, then v is PS.
15. If e is Z and $\dot{e}$ is PL, then v is PL.
16. If e is PS and $\dot{e}$ is NL, then v is NS.
17. If e is PS and $\dot{e}$ is NS, then v is Z.
18. If e is PS and $\dot{e}$ is Z, then v is PS.
19. If e is PS and $\dot{e}$ is PS, then v is PL.
20. If e is PS and $\dot{e}$ is PL, then v is PL.
21. If e is PL and $\dot{e}$ is NL, then v is Z.
22. If e is PL and $\dot{e}$ is NS, then v is PS.
23. If e is PL and $\dot{e}$ is Z, then v is PL.
24. If e is PL and $\dot{e}$ is PS, then v is PL.
25. If e is PL and $\dot{e}$ is PL, then v is PL.

These are given in tabular form (Table 4.1) as follows:

TABLE 4.1 Tabulated Rule Base for Ball and Beam Controller

	v	$\dot{e}$				
		NL	NS	Z	PS	PL
	NL	NL	NL	NL	NS	Z
	NS	NL	NL	NS	Z	PS
e	Z	NL	NS	Z	PS	PL
	PS	NS	Z	PS	PL	PL
	PL	Z	PS	PL	PL	PL

For later reference, let us define the *rule matrix* of this controller as

$$V_{e\dot{e}} = \begin{bmatrix} NL & NL & NL & NS & Z \\ NL & NL & NS & Z & PS \\ NL & NS & Z & PS & PL \\ NS & Z & PS & PL & PL \\ Z & PS & PL & PL & PL \end{bmatrix} \tag{4.2}$$

4.1.3 Inference: Ball and Beam Plant

Since *product* T-norm is used (see Section 3.6.5) the premise value of each rule at each time t is the product of the degrees of membership of $e(t)$ in the fuzzy set for e specified by the rule and $\dot{e}(t)$ in the fuzzy set for $\dot{e}$ specified by the rule. For instance, the premise value for Rule 1 at time t is

$$\mu_1(e(t), \dot{e}(t)) = \mu_e^{NL}(e(t))\mu_{\dot{e}}^{NL}(\dot{e}(t)) \tag{4.3}$$

Consider a particular time t at which $[e(t), \dot{e}(t))] = (-0.0625\,\text{m}, 3\,\text{m/s})$. This situation corresponds to the ball being slightly to the right of center and traveling rapidly to the left. Referring to Figure 4.3, a tracking error of $e = -0.0625$ is considered Z to an extent 0.75 and NS to an extent 0.25 (this e is not in any of the other three fuzzy sets on the e universe). Similarly, referring to Figure 4.4, $\dot{e} = 3$ is considered PL to an extent 0.5 and PS to an extent 0.5. Therefore, we have

$$\mu_e^{Z}(-0.0625) = 0.75 \tag{4.4a}$$
$$\mu_e^{NS}(-0.0625) = 0.25 \tag{4.4b}$$
$$\mu_{\dot{e}}^{PS}(3) = 0.5 \tag{4.4c}$$
$$\mu_{\dot{e}}^{PL}(3) = 0.5 \tag{4.4d}$$

Rules 9, 10, 14, and 15 are *on*, and the rest are not fired.

The premise values of the fired rules are

$$\mu_9(-0.0625, 3) = \mu_e^{NS}(-0.0625)\mu_{\dot{e}}^{PS}(3) = 0.25(0.5) = 0.125 \tag{4.5a}$$
$$\mu_{10}(-0.0625, 3) = \mu_e^{NS}(-0.0625)\mu_{\dot{e}}^{PL}(3) = 0.25(0.5) = 0.125 \tag{4.5b}$$
$$\mu_{14}(-0.0625, 3) = \mu_e^{Z}(-0.0625)\mu_{\dot{e}}^{PS}(3) = 0.75(0.5) = 0.375 \tag{4.5c}$$
$$\mu_{15}(-0.0625, 3) = \mu_e^{Z}(-0.0625)\mu_{\dot{e}}^{PL}(3) = 0.75(0.5) = 0.375 \tag{4.5d}$$

4.1.4 Defuzzification: Ball and Beam Plant

Since the output memberships are singletons (see Section 3.6.5), the crisp output of the fuzzy controller at time t, which is the voltage input to the motor at time t, is calculated using center average defuzzification as

$$v(t)=\frac{\sum_{i=1}^{25} q_i\mu_i(e(t),\dot{e}(t))}{\sum_{i=1}^{25} \mu_i(e(t),\dot{e}(t))}=\frac{q_9\mu_9+q_{10}\mu_{10}+q_{14}\mu_{14}+q_{15}\mu_{15}}{\mu_9+\mu_{10}+\mu_{14}+\mu_{15}} \tag{4.6}$$

where q_i is the location of the membership function characterizing the singleton fuzzy set specified in the consequent of Rule i. The crisp output of the fuzzy controller for the inputs $[e(t), \dot{e}(t)] = (-0.0625, 3)$ is

$$v(t)=\frac{0(0.125)+5(0.125)+5(0.375)+10(0.375)}{0.125+0.125+0.375+0.375}=6.25\ V \tag{4.7}$$

Thus the controller commands a clockwise rotation in order to stop the ball's leftward motion. Note that the denominator in (4.7) is unity due to the partitions of unity on the e and $\dot{e}$ universes (Figs. 4.3 and 4.4).

4.2 TUNING FOR IMPROVED PERFORMANCE BY ADJUSTING SCALING GAINS

As is done in many control systems both fuzzy and nonfuzzy, let us add adjustable scaling gains g_0 and g_1 for inputs e and $\dot{e}$, respectively, and adjustable scaling gain h for output v. These gains are used to tune the compensator to achieve better performance. The closed-loop system is shown in Figure 4.9, where the contents of the fuzzy compensator block are shown in Figure 4.10.

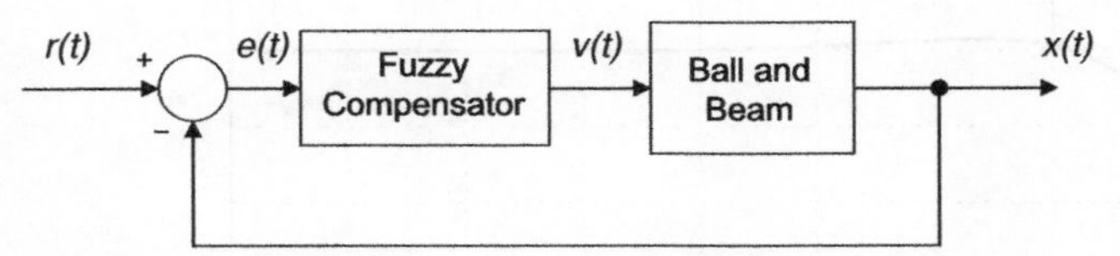

Figure 4.9. Closed-loop controller for ball and beam.

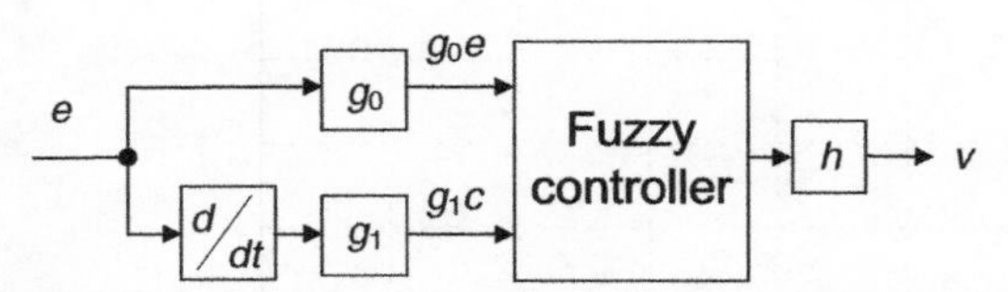

Figure 4.10. Fuzzy compensator block of Figure 4.9.

When the closed-loop system is simulated using a fourth-order Runge–Kutta integration routine with a step size of $\Delta t = 0.001$ s and an initial ball position of

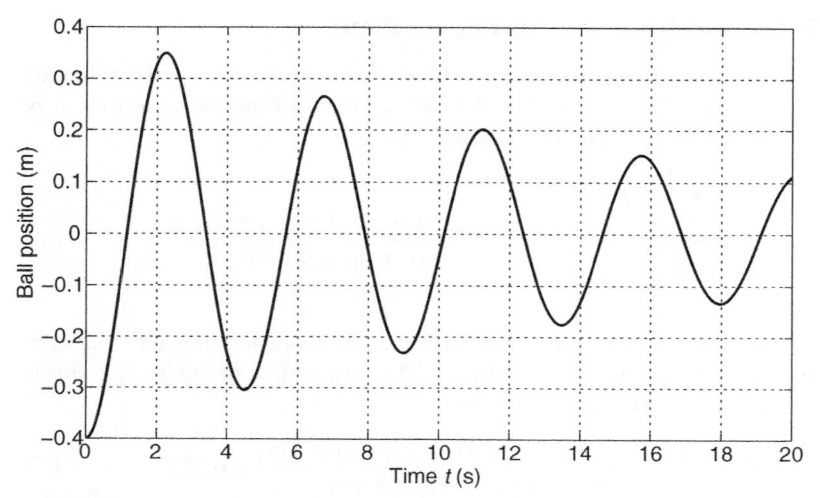

Figure 4.11. Ball travel, scaling gains $g_0 = 1$, $g_1 = 1$, $h = 1$.

$x(0) = -0.4$ m, the resulting ball movement is shown in Figure 4.11. The ball moves toward zero, however, the response is very oscillatory. The fuzzy controller is doing its job, but must be tuned to quicken the response and get rid of the oscillations. This can be done by adjusting the gains g_0, g_1, and h.

In some cases, the fuzzy controller originally designed with all scaling gains equaling unity may not produce a stable closed-loop system, even though the controller is correctly designed according to our best judgment. In such cases, the problem is likely that the universes of discourse are incorrectly sized for the problem. Then it may be necessary to adjust the scaling gains initially to produce a stable closed-loop system before the gains are further adjusted to improve the response.

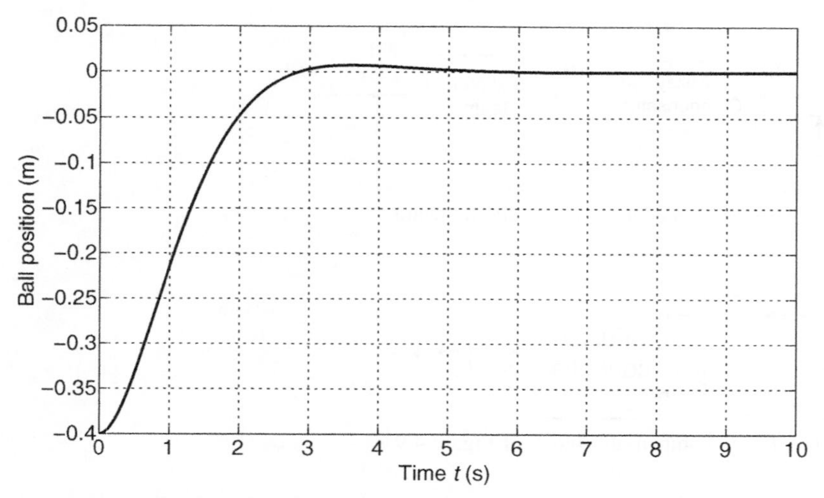

Figure 4.12. Ball travel, scaling gains $g_0 = 1$, $g_1 = 18$, $h = 1$.

When g_1 is increased to 18 to get rid of the oscillations, the behavior shown in Figure 4.12 results. This reduces the oscillations significantly, but the ball takes 6 s to settle motionless at the center of the beam, which is too slow.

In order to quicken the response, g_0 is increased to 3 while keeping $g_1 = 18$. The resulting response is shown in Figure 4.13. The time taken by the ball to reach the center of the beam is significantly decreased, however, the ball overshoots its target.

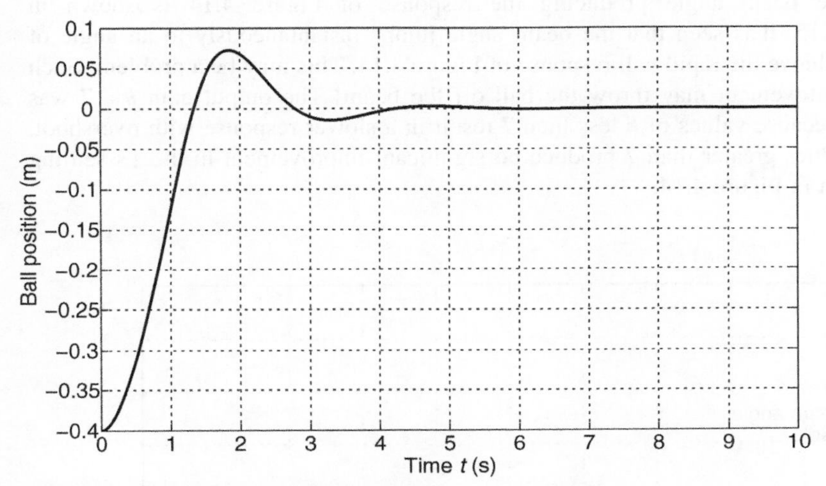

Figure 4.13. Ball travel, scaling gains $g_0 = 3$, $g_1 = 18$, $h = 1$.

When the output scaling gain h is increased to 7 while keeping g_0 and g_1 unchanged, the response of Figure 4.14 results.

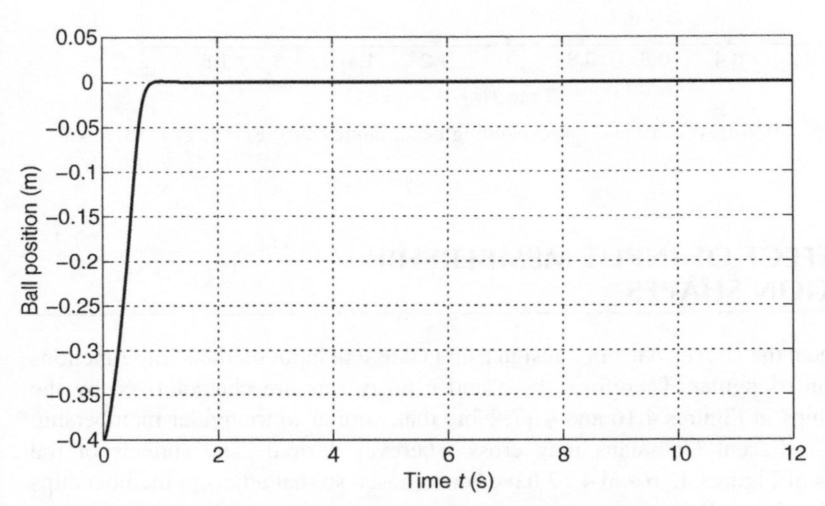

Figure 4.14. Ball travel, scaling gains $g_0 = 3$, $g_1 = 18$, $h = 7$.

This is a satisfactory response; the ball settles at the center of the beam within 1 s with no overshoot. However, note that, with $h = 7$, the controller now requires that the voltage source be able to supply voltages in the range $-70 \leq v \leq 70\,V$ at currents demanded by the motor. If a sufficiently powerful supply is not available, it may be necessary to accept a slower response. The control effort necessary to accomplish a control task is always a concern in the implementation of every practical control system.

The beam angle producing the response of Figure 4.14 is shown in Figure 4.15. It is seen that the beam angle jumps instantaneously to an angle of 40° to achieve the rapid ball response of Figure 4.14. This may be a problem: Such a quick movement may throw the ball off the beam! The output gain $h = 7$ was chosen because values of h less then 7 result in a slower response with overshoot, while values greater than 7 produce no significant improvement in the 1 s settling time seen in Figure 4.14.

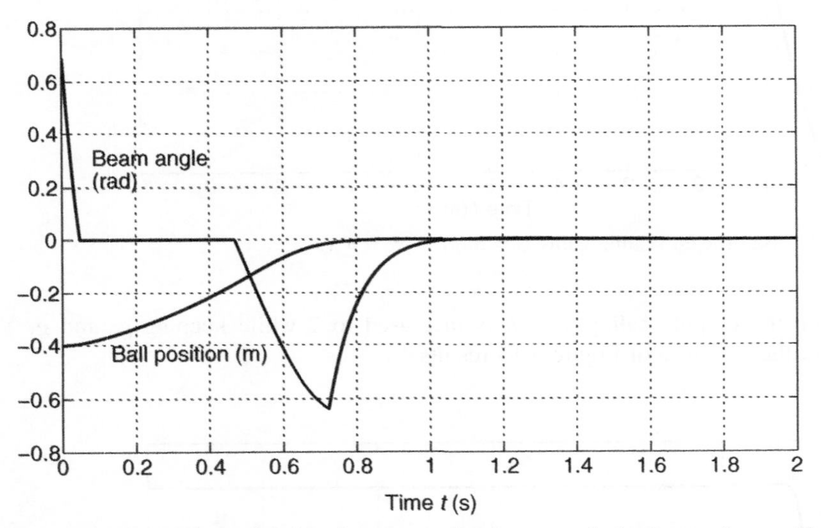

Figure 4.15. Ball travel (m) and corresponding beam angle (rad), $g_0 = 3$, $g_1 = 18$, $h = 7$.

4.3 EFFECT OF INPUT MEMBERSHIP FUNCTION SHAPES

Let us repeat the above controller design using Gaussian input membership functions rather than triangular. Therefore, the e and $\dot{e}$ fuzzy sets are characterized by the memberships in Figures 4.16 and 4.17. Note that, similar to triangular membership functions, adjacent Gaussians may cross wherever desired. The spreads of the Gaussians of Figures 4.16 and 4.17 have been chosen so that adjacent memberships cross each other at 0.5.

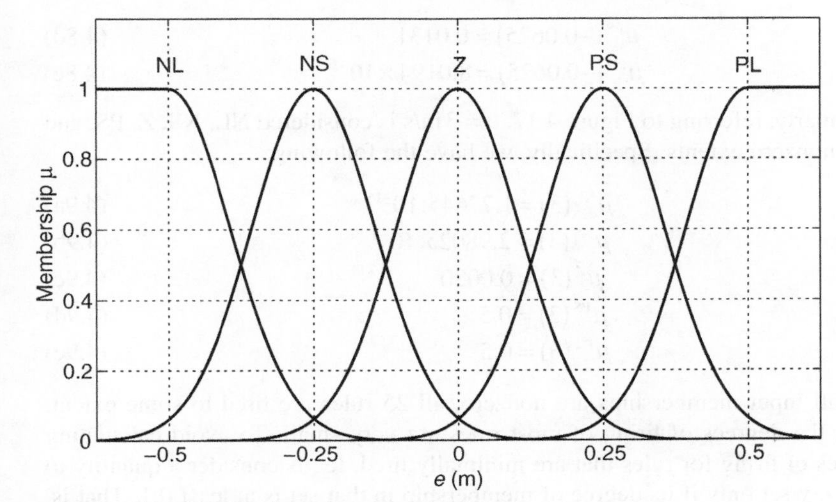

Figure 4.16. Gaussian fuzzy sets on e universe.

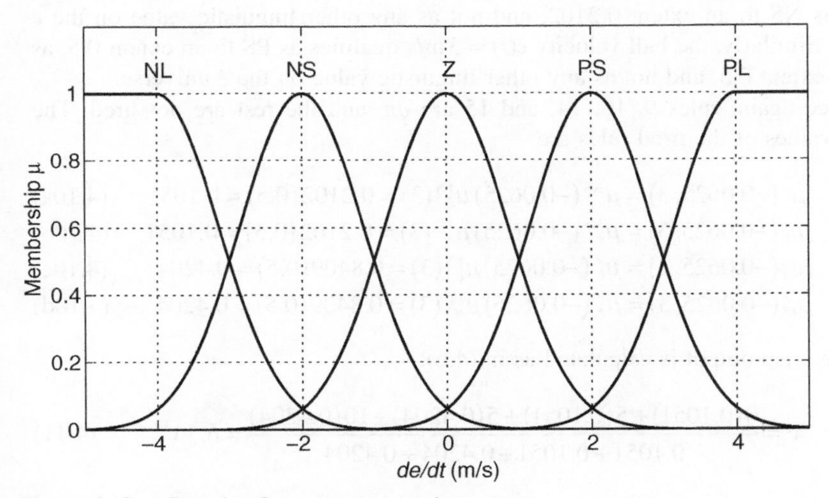

Figure 4.17. Gaussian fuzzy sets on $\dot{e}$ universe.

The rule base of Section 4.1.2 has not changed. However, the inference calculation is different due to the different membership function shapes. As above, consider a particular time t at which $[e(t), \dot{e}(t)] = (-0.0625\,\text{m}, 3\,\text{m/s})$. Referring to Figure 4.16, a tracking error of $e = -0.0625\,\text{m}$ is considered NL, NS, Z, PS, and PL all to nonzero extents. Specifically, we have the following:

$$\mu_e^{\text{NL}}(-0.0625) = 2.0530\times10^{-4} \tag{4.8a}$$

$$\mu_e^{\text{NS}}(-0.0625) = 0.2102 \tag{4.8b}$$

$$\mu_e^{\text{Z}}(-0.0625) = 0.8409 \tag{4.8c}$$

$$\mu_e^{\text{PS}}(-0.0625) = 0.0131 \tag{4.8d}$$
$$\mu_e^{\text{PL}}(-0.0625) = 8.0194 \times 10^{-7} \tag{4.8e}$$

Similarly, referring to Figure 4.17, $\dot{e} = 3\,\text{m/s}$ is considered NL, NS, Z, PS, and PL all to nonzero extents. Specifically, we have the following:

$$\mu_{\dot{e}}^{\text{NL}}(3) = 1.7764 \times 10^{-15} \tag{4.9a}$$
$$\mu_{\dot{e}}^{\text{NS}}(3) = 2.9802 \times 10^{-8} \tag{4.9b}$$
$$\mu_{\dot{e}}^{\text{Z}}(3) = 0.0020 \tag{4.9c}$$
$$\mu_{\dot{e}}^{\text{PS}}(3) = 0.5 \tag{4.9d}$$
$$\mu_{\dot{e}}^{\text{PL}}(3) = 0.5 \tag{4.9e}$$

Because all input memberships are nonzero, all 25 rules are fired to some extent. However, the degrees of firing of most rules are very small. To avoid calculating the degrees of firing for rules that are minimally fired, let us consider a quantity to be in a fuzzy set only if its degree of membership in that set is at least 0.1. That is, we consider a quantity to be in a fuzzy set only if it is in the 0.1-cut of that set. Using this criterion, the ball position $e(t) = -0.0625\,\text{m}$ qualifies as Z to an extent 0.8409, as NS to an extent 0.2102, and not as any other linguistic value on the e universe. Similarly, the ball velocity $\dot{e}(t) = 3\,\text{m/s}$ qualifies as PS to an extent 0.5, as PL to an extent 0.5, and not as any other linguistic value on the $\dot{e}$ universe.

Once again, rules 9, 10, 14, and 15 are *on*, and the rest are not fired. The premise values of the fired rules are

$$\mu_9(-0.0625, 3) = \mu_e^{\text{NS}}(-0.0625)\mu_{\dot{e}}^{\text{PS}}(3) = 0.2102(0.5) = 0.1051 \tag{4.10a}$$
$$\mu_{10}(-0.0625, 3) = \mu_e^{\text{NS}}(-0.0625)\mu_{\dot{e}}^{\text{PL}}(3) = 0.2102(0.5) = 0.1051 \tag{4.10b}$$
$$\mu_{14}(-0.0625, 3) = \mu_e^{\text{Z}}(-0.0625)\mu_{\dot{e}}^{\text{PS}}(3) = 0.8409(0.5) = 0.4204 \tag{4.10c}$$
$$\mu_{15}(-0.0625, 3) = \mu_e^{\text{Z}}(-0.0625)\mu_{\dot{e}}^{\text{PL}}(3) = 0.8409(0.5) = 0.4204 \tag{4.10d}$$

The crisp output is calculated as in (4.6):

$$v = \frac{0(0.1051) + 5(0.1051) + 5(0.4204) + 10(0.4204)}{0.1051 + 0.1051 + 0.4204 + 0.4204} = 6.5\,V \tag{4.11}$$

This is similar to the crisp output calculated in (4.7) for triangular memberships. In fact, the closed-loop behavior of the plant with Gaussian controller is very close to that of the plant with triangular controller. This is not surprising, because the shape of Gaussian memberships is similar to that of triangular memberships. Neither controller is "better." The difference is analogous to the difference that occurs when two different people try to balance the ball on the beam by hand.

The input–output characteristic of the fuzzy controller with triangular memberships is shown in Figure 4.18, and that of the controller with Gaussian memberships is shown in Figure 4.19. The two characteristics are essentially the same, except for the local waves and undulations in the characteristic of Figure 4.19. These were discussed in Section 3.6.1. These local waves are due to the curvature of the Gaussians; they have nothing to do with the controller itself.

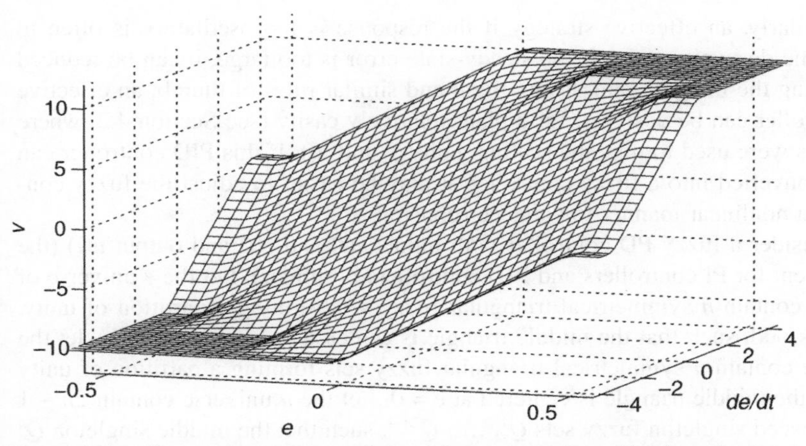

Figure 4.18. Input–output characteristic of fuzzy controller with triangular input memberships, *product* T-norm, and singleton output fuzzy sets.

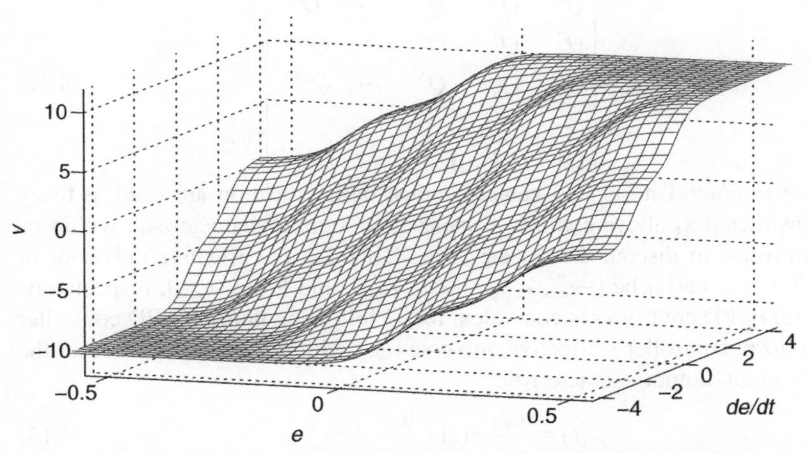

Figure 4.19. Input–output characteristic of fuzzy controller with Gaussian input memberships, *product* T-norm, and singleton output fuzzy sets.

4.4 CONVERSION OF PID CONTROLLERS INTO FUZZY CONTROLLERS

There is an exact correspondence between PID controllers and certain fuzzy controllers. This makes it possible to derive a fuzzy controller from a PID controller. It is very useful to be able to do this because designing a fuzzy controller can be initially difficult (as was seen in Sections 4.1 and 4.2). However, there exist well-known methods of designing PID controllers (e.g., Ziegler–Nichols tuning [3]).

Even when these methods are not used, an expert can often manually adjust a PID controller to achieve stable performance. For instance, in many applications if the response is too sluggish, it can be quickened by increasing the proportional gain. Similarly, an effective strategy if the response is too oscillatory is often to increase the derivative gain. If the steady-state error is too large, it can be reduced by adjusting the integral gain. Using these and similar rules of thumb, an effective PID controller can be designed for a plant relatively easily (see Section 4.2, where these rules were used to tune the scaling gains g_0 and g_1). If this PID controller can then be converted into a fuzzy controller, it may be easier to adjust the fuzzy controller in a nonlinear manner to enhance robustness.

Consider a fuzzy PD controller with inputs $e(t)$ and $\dot{e}(t)$ and output $u(t)$ (the development for PI controllers and PID controllers is similar). Let the e universe of discourse contain n symmetrical triangular fuzzy sets forming a partition of unity, where n is odd, such that the middle triangle is centered at $e = 0$. Similarly, let the $\dot{e}$ universe contain n symmetrical triangular fuzzy sets forming a partition of unity such that the middle triangle is centered at $\dot{e} = 0$. Let the u universe contain $2n - 1$ equally spaced singleton fuzzy sets $Q^1, \ldots, Q^{2n-1}$, such that the middle singleton Q^n is located at $u = 0$. Finally, let the rule base of the controller be specified by the $n \times n$ rule matrix [see (4.2)]

$$U_{e\dot{e}} = \begin{bmatrix} Q^1 & Q^2 & Q^3 & \cdots & Q^n \\ Q^2 & Q^3 & Q^4 & \cdots & Q^{n+1} \\ Q^3 & Q^4 & Q^5 & \cdots & Q^{n+2} \\ \vdots & \vdots & \vdots & \ddots & \vdots \\ Q^n & Q^{n+1} & Q^{n+2} & & Q^{2n-1} \end{bmatrix} \tag{4.12}$$

When *product* T-norm and center-average defuzzification are used, a fuzzy system constructed as above exhibits a linear input–output characteristic within its effective universe of discourse (see, e.g., Fig. 4.18). Let the effective universes of discourse for e, $\dot{e}$, and u be $[-e_{\max}\ e_{\max}]$, $[-\dot{e}_{\max}\ \dot{e}_{\max}]$, and $[-u_{\max}\ u_{\max}]$, respectively. Then, the fuzzy PD controller is equivalent to the following nonfuzzy PD controller within the fuzzy controller's effective universe (i.e., within the linear portion of the controller's input–output characteristic):

$$u(t) = \frac{u_{\max}}{e_{\max}} e(t) + \frac{u_{\max}}{\dot{e}_{\max}} \dot{e}(t) \tag{4.13}$$

Therefore, the fuzzy controller whose characteristic is shown in Figure 4.18 is equivalent to the nonfuzzy PD controller

$$u(t) = \frac{10}{0.5} e(t) + \frac{10}{4} \dot{e}(t) = 20e(t) + 2.5\dot{e}(t) \tag{4.14}$$

This suggests a method for deriving a fuzzy controller from a nonfuzzy PD controller. Assume that we have a nonfuzzy PD controller that gives satisfactory closed-loop behavior. In order for the equivalent fuzzy controller to be exactly equal to the nonfuzzy PD, it is necessary that the system trajectory always remain within the linear part of the characteristic. Therefore, with the nonfuzzy PD in operation, measure the maximum control effort necessary to accomplish the control task. This is the maximum absolute value of $u(t)$ that is output by the PD in controlling the plant. Call this $u_{\max}$. Then if the nonfuzzy PD controller is given by

$$u(t) = K_p e(t) + K_d \dot{e}(t) \tag{4.15}$$

the equivalent fuzzy controller has inputs $e(t)$, $\dot{e}(t)$, and output $u(t)$ with effective universes $\left[-\frac{\gamma u_{\max}}{K_p}, \frac{\gamma u_{\max}}{K_p}\right]$, $\left[-\frac{\gamma u_{\max}}{K_d}, \frac{\gamma u_{\max}}{K_d}\right]$, and $[-2\gamma u_{\max} \quad 2\gamma u_{\max}]$, respectively. The constant $\gamma > 1$ is to guarantee the system trajectories remain within the linear portion of the fuzzy system's input–output characteristic. Usually $\gamma = 2$ will suffice, but any sufficiently large γ will produce identical closed-loop behavior to the non-fuzzy PD compensator. After the equivalent fuzzy controller has been constructed, it can be easily altered to improve performance.

EXAMPLE 4.1

Consider the inverted pendulum plant of Figure 1.2. For simulation purposes, its mathematical model is given by

$$\ddot{\psi} = \frac{9.81\sin\psi - \frac{2}{3}\cos\psi\left(0.25\dot{\psi}^2\sin\psi + F\right)}{0.5\left(\frac{4}{3} - \frac{1}{3}\cos^2\psi\right)} \tag{4.16}$$

The error e is defined as $e = r(t) - \psi(t)$, where $\psi(t)$ is the angle of the rod from vertical and $r(t)$ is a reference trajectory. Suppose it has been determined that the PD controller

$$F(t) = -(30e(t) + 5\dot{e}(t)) \tag{4.17}$$

quickly balances the pendulum in the vertical-up position with no overshoot. Figures 4.20 and 4.21 show the rod angle response and corresponding PD controller output, respectively for an initial rod angle of –0.1 rad.

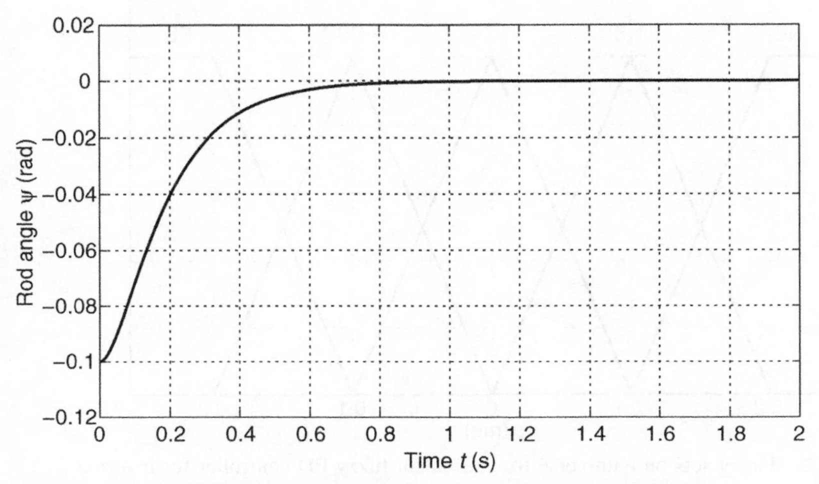

Figure 4.20. Rod angle response under nonfuzzy PD control (4.17).

From Figure 4.21, the maximum absolute value of the control effort is 3N. If we use $n = 5$ fuzzy sets for $e(t)$ and $\dot{e}(t)$, the input and output universes given in Figures 4.22–4.24 result.

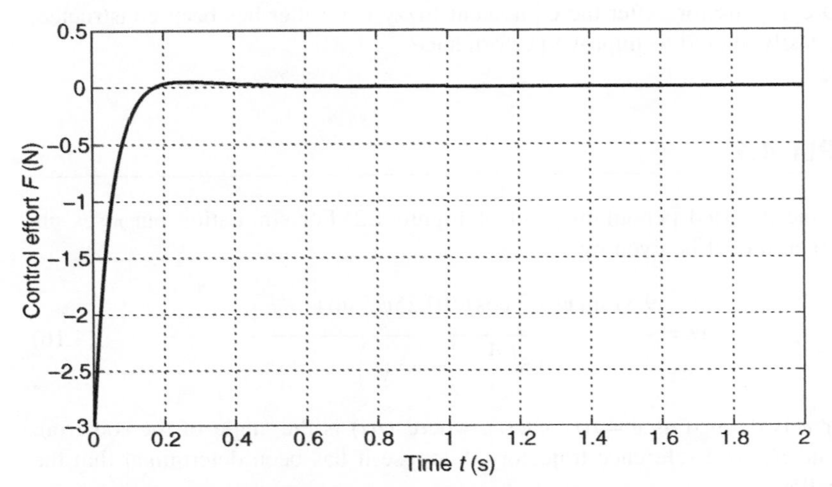

Figure 4.21. Control effort [output of nonfuzzy PD controller (4.17)] producing response of Figure 4.20.

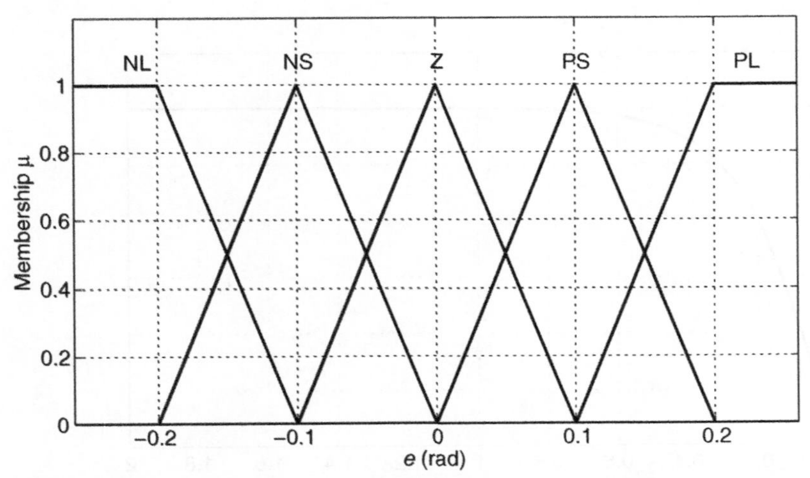

Figure 4.22. Fuzzy sets on e universe for equivalent fuzzy PD controller for inverted pendulum.

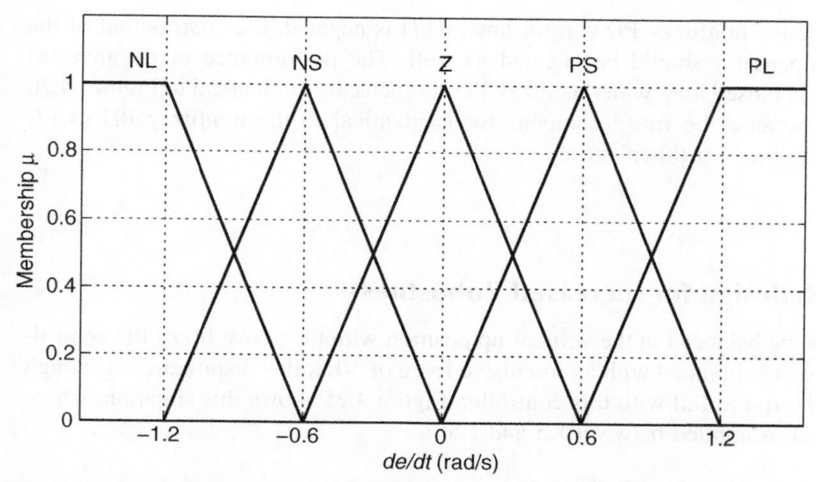

Figure 4.23. Fuzzy sets on $\dot{e}$ universe for equivalent fuzzy PD controller for inverted pendulum.

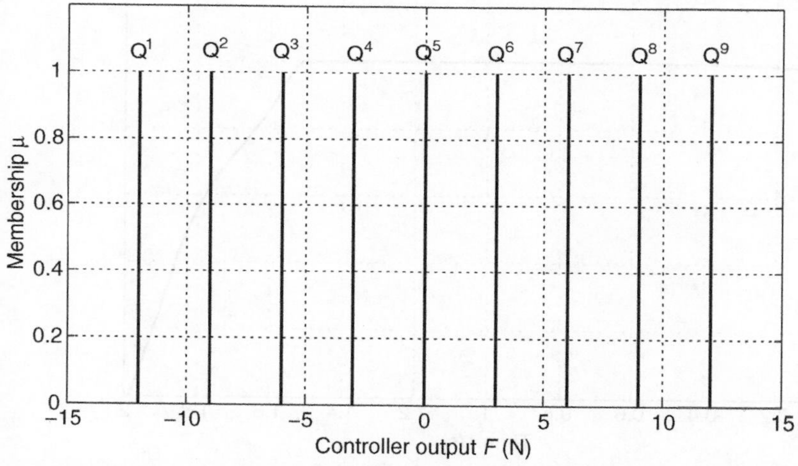

Figure 4.24. Fuzzy sets on F universe for equivalent fuzzy PD controller for inverted pendulum.

The tabulated rule base of the controller is, from (4.12),

TABLE 4.2 Tabulated Rule Base for Inverted Pendulum Controller

F		$\dot{e}$				
		NL	NS	Z	PS	PL
e	NL	Q^1	Q^2	Q^3	Q^4	Q^5
	NS	Q^2	Q^3	Q^4	Q^5	Q^6
	Z	Q^3	Q^4	Q^5	Q^6	Q^7
	PS	Q^4	Q^5	Q^6	Q^7	Q^8
	PL	Q^5	Q^6	Q^7	Q^8	Q^9

Since the nonfuzzy PD control law (4.17) is negated, the crisp output of the fuzzy compensator should be negated as well. The performance of the inverted pendulum in closed loop with this fuzzy PD compensator is identical to Figures 4.20 and 4.21 because the fuzzy compensator is identical to the nonfuzzy PD (4.17) initially designed for this plant.

□

4.4.1 Redesign for Increased Robustness

Let the rod be balanced in the vertical-up position with the above fuzzy PD controller. If the cart is bumped with an impulsive force of 50 N, this disturbance is enough to cause the rod to fall with this controller. Figure 4.25 shows this situation, where a 50 N force is applied between 1.5 and 1.55 s.

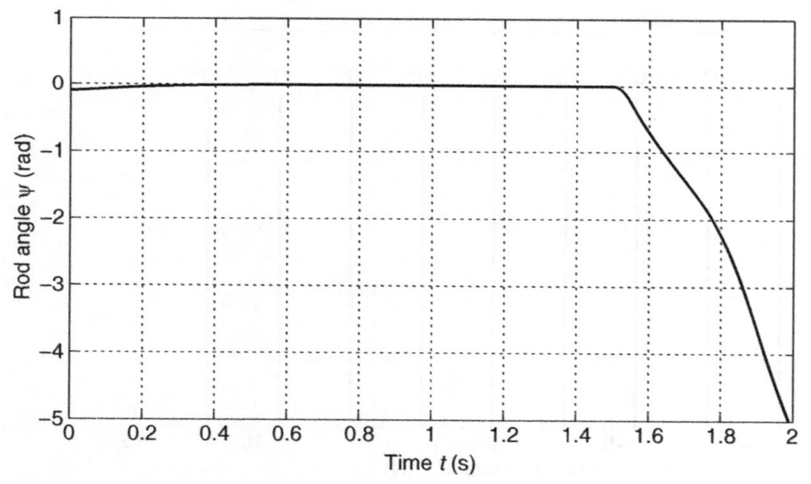

Figure 4.25. Rod angle, 50 N impulsive disturbance at $t = 1.5$ s.

The fuzzy PD can easily be redesigned on the basis of expert knowledge so that the rod's balance will not be destroyed by such impulsive disturbances. The redesign consists of repositioning the output singletons so that the controller delivers greater force when e and $\dot{e}$ are large, but when e and $\dot{e}$ are small the control is unchanged. Therefore, let the new output singletons be as in Figure 4.26.

The input–output characteristic of the redesigned nonlinear fuzzy PD controller is shown in Figure 4.27. In this figure, the characteristic of the redesigned nonlinear fuzzy PD is the same as that of the linear controller when small control

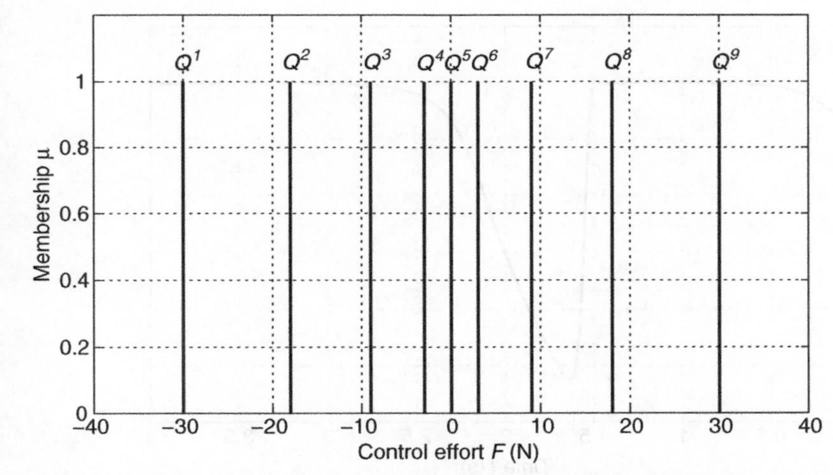

Figure 4.26. Redesigned output singletons for fuzzy PD controller for inverted pendulum.

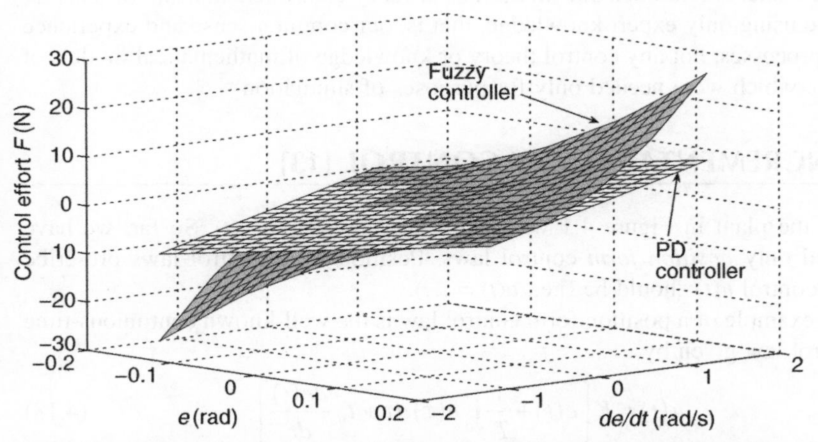

Figure 4.27. Input–output characteristics of linear PD and redesigned nonlinear fuzzy controllers.

effort is required, but for cases when large control effort is required the redesigned fuzzy controller differs significantly from the linear controller.

The response of the inverted pendulum under closed-loop control with the redesigned nonlinear fuzzy controller is shown in Figure 4.28. In this figure, we see that now the cart catches the rod before it falls, unlike Figure 4.25. Note that the fuzzy PD was easily redesigned using expert knowledge to handle the impulsive disturbance. The original nonfuzzy PD (4.17) may not be so easily redesigned to increase robustness.

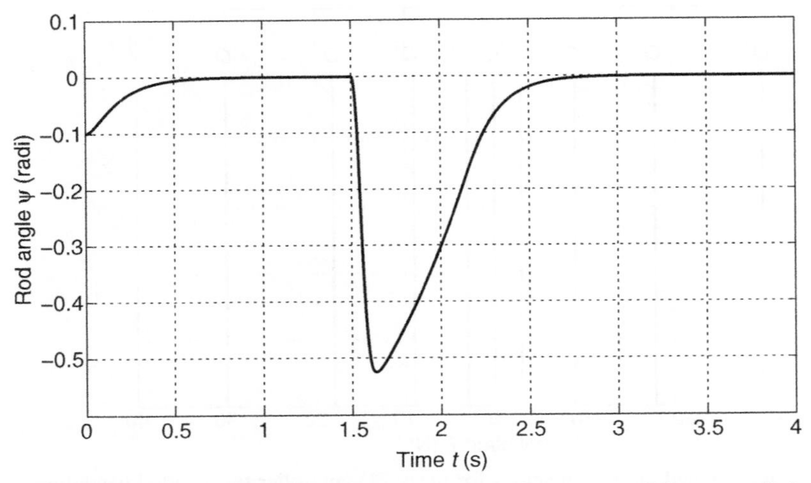

Figure 4.28. Rod angle with 50 N impulsive disturbance at t = 1.5 s, controlled with redesigned (nonlinear) fuzzy PD controller.

The reader is reminded that all designs of fuzzy controllers done in this chapter were done using only expert knowledge, that is, our common sense and experience with the processes, not any control theory or knowledge of mathematical models of the plants, which were needed only for purposes of simulation.

4.5 INCREMENTAL FUZZY CONTROL [13]

Consider the plant in Figure 4.1 with input $u(t)$ and output $y(t)$. So far, we have considered only *position form* control laws. Position form control laws prescribe what the control $u(t)$ should be (i.e., $u(t) = \ldots$).

An example of a position form control law is the well-known continuous-time PID control law given by

$$u(t) = K\left[e(t) + \frac{1}{T_i}\int_0^t e(\tau)\,d\tau + T_d\frac{de(t)}{dt}\right] \tag{4.18}$$

where $e(t)$ is the summer output and K, T_i, and T_d are constants chosen by the designer. Let $r(t) = r$, a constant. If we sample $e(t)$ every Δt seconds, we obtain the discrete-time position form PID control law given by

$$\begin{aligned} u(k) &= K\left[e(k) + \frac{\Delta t}{T_i}\sum_{i=0}^{k} e(i) + \frac{T_d}{\Delta t}c(k)\right] = Ke(k) + \frac{K\Delta t}{T_i}\sum_{i=0}^{k} e(i) + \frac{KT_d}{\Delta t}c(k) \\ &= K_p e(k) + K_i\sum_{i=0}^{k} e(i) + K_d c(k) \end{aligned} \tag{4.19}$$

where $c(k) = e(k) - e(k-1)$.

In *incremental form* control laws, the *change* in plant input, rather than the input itself, is prescribed by the controller. The change in input is given by $\Delta u(k) = u(k) - u(k-1)$. To formulate an incremental PID control law, we need [from (4.19)]

$$u(k-1) = K_p e(k-1) + K_i \sum_{i=0}^{k-1} e(i) + K_d c(k-1) \tag{4.20}$$

Therefore, the resulting PID control law in incremental form is given by

$$\Delta u(k) = K_p c(k) + K_i e(k) + K_d d(k) \tag{4.21}$$

where $d(k) = c(k) - c(k-1)$. Then, the control delivered to the plant is

$$u(k) = u(k-1) + \Delta u(k) \tag{4.22}$$

EXAMPLE 4.2

An incremental fuzzy controller is designed for the ball and beam of Section 4.1 using the method of Section 4.4 with $\Delta t = 0.01$ s. First, a nonfuzzy PD incremental controller is designed in an ad hoc manner using basic knowledge of PID controllers (increasing proportional gain quickens the response, increasing derivative gain decreases oscillations, etc.):

$$\Delta v(k) = P_{\text{incr}} c(k) + D_{\text{incr}} d(k) \tag{4.23}$$

This trial and error method results in gains of $P_{\text{incr}} = 500$, $D_{\text{incr}} = 58000$ to approximately duplicate the closed-loop performance of the ball and beam with position-form PD controller (4.17).

When the plant is run in closed loop with this controller, the maximum controller output is measured as $\Delta v_{\max} = 211.65$ V. Therefore, letting $\gamma = 1.89$ (for round numbers for locations of output singletons) and assuming there are three fuzzy sets on each universe, the effective universes of discourse for c, d, and Δv are, respectively, $-0.8 \le c \le 0.8$, $-0.0069 \le d \le 0.0069$, and $-800 \le \Delta v \le 800$, resulting in the following fuzzy sets on these universes (Figs. 4.29–4.31):

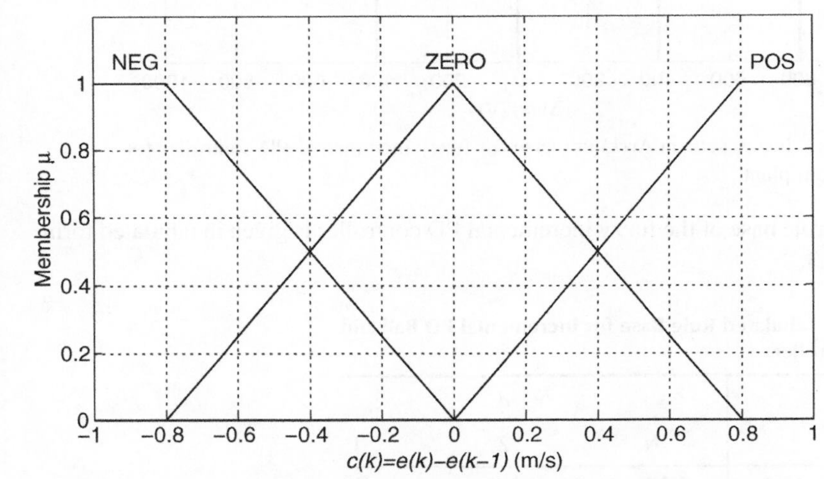

Figure 4.29. Fuzzy sets on $c(k)$ universe for fuzzy incremental PD controller for ball and beam plant.

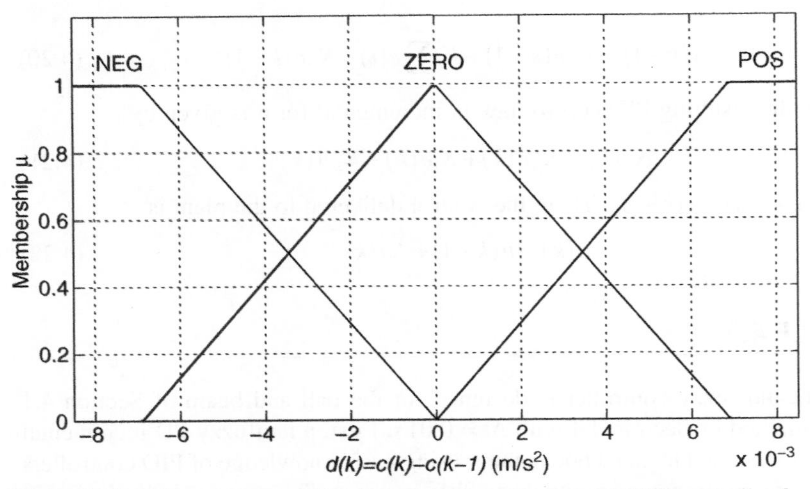

Figure 4.30. Fuzzy sets on $d(k)$ universe for fuzzy incremental PD controller for ball and beam plant.

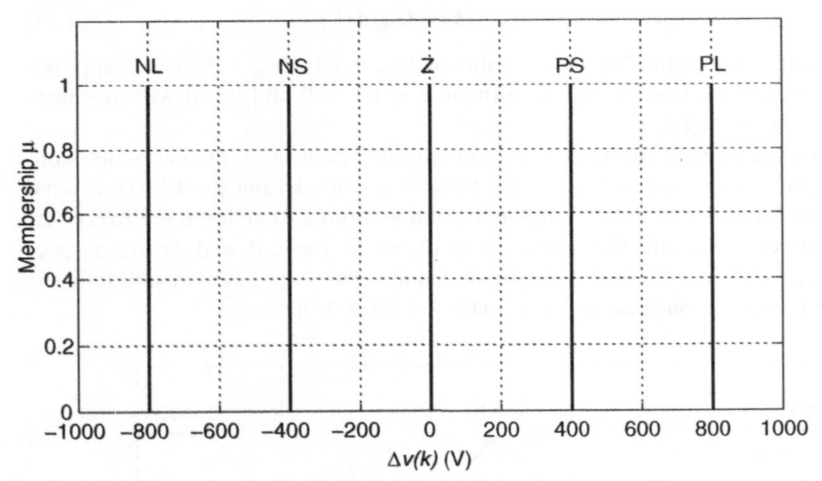

Figure 4.31. Fuzzy sets on $\Delta v(k)$ universe for fuzzy incremental PD controller for ball and beam plant.

The rule base of the fuzzy incremental PD controller is given in tabulated form in Table 4.3.

□

TABLE 4.3 Tabulated Rule Base for Incremental PD Ball and Beam Controller

Δv		d		
		N	Z	P
c	N	NL	NS	Z
	Z	NS	Z	PS
	P	Z	PS	PL

4.6 SUMMARY

Mamdani fuzzy systems have fuzzy sets in their rule consequents, as opposed to Takagi–Sugeno fuzzy systems (see Chapter 6), which have mathematical expressions in their rule consequents. We consider the most common interconnection for tracking: unity feedback with a fuzzy controller in casacade with the plant. Fuzzy control is not **model based**, that is, its design does not depend on a mathematical model of the plant, but on the designer's common sense and past experience with the process being controlled.

For the **ball and beam** controller (Sections 4.1–4.3), we first decide on the inputs that would be needed (e and $\dot{e}$), then determine proper universes of discourse for all inputs and outputs. These are based on our best common sense about the system, but scaling gains are included in the design to improve performance.

Section 4.4 introduces a method for **conversion of PID controllers into fuzzy controllers**. The equivalent fuzzy controller is exactly equal to the PID within its effective universe. This technique is useful because nonfuzzy PID controllers can be readily designed using well-known techniques. Once the equivalent fuzzy controller has been obtained, it can be easily adjusted in a nonlinear manner using heuristic knowledge to improve performance. Example 4.1 shows how this can be done. In this example, a nonfuzzy PD controller for an inverted pendulum plant is converted to a fuzzy controller. This fuzzy controller is then adjusted in a nonlinear manner to increase robustness of the system to impulsive disturbances.

Section 4.5 introduces **incremental fuzzy control**, which prescribes the *change* in input rather than the input itself. A fuzzy incremental PID is derived, requiring inputs $c(k)$, $e(k)$, and $d(k)$ where $e(k) = r - y(k)$, $c(k) = e(k) - e(k-1)$, and $d(k) = c(k) - c(k-1)$. In Example 4.2, the method of Section 4.4 is used to derive an incremental fuzzy PD controller for the ball and beam plant. This controller approximately duplicates the closed-loop performance of the position-form controller derived in Section 4.1.

EXERCISES

4.1 In the fuzzy controller of Section 4.1, calculate the crisp output for an input of (a) $(e, \dot{e}) = (-0.3, 0.25)$; (b) $(e, \dot{e}) = (-0.0625, 0.25)$.

4.2 Design a fuzzy controller for the ball and beam similar to the one in Section 4.1, but use 7 triangular fuzzy sets on each universe instead of 5.

4.3 Repeat Problem 4.1 for the controller of Section 4.3.

4.4 Design a fuzzy controller for the ball and beam similar to the one in Section 4.3, but use three Gaussian fuzzy sets on each universe instead of 5.

4.5 Derive a fuzzy controller equivalent to the PID controller (4.18) with $K = 10$, $T_i = 1$, and $T_d = 5$. The maximum control effort from the PID is 3.

4.6 Derive (4.13).

4.7 Plot the input–output characteristic of the PD controller $u(t) = K_p e(t) + K_d \dot{e}(t)$ where $K_p = K_d = 1$.

4.8 Derive the relationship between a PID and a fuzzy controller in Section 4.4, that is, that the nonfuzzy PD controller (4.15) is equivalent to a fuzzy controller with inputs $e(t)$ and $\dot{e}(t)$ such that the effective universe for $e(t)$ is $\left[-\frac{\gamma u_{\max}}{K_p}, \frac{\gamma u_{\max}}{K_p}\right]$, the effective universe for $\dot{e}(t)$ is $\left[-\frac{\gamma u_{\max}}{K_d}, \frac{\gamma u_{\max}}{K_d}\right]$, and the effective universe for $u(t)$ is $[-2\gamma u_{\max} \quad 2\gamma u_{\max}]$ where $u_{\max}$ is the maximum control effort required for the control task and γ is a positive constant.

4.9 Using simulation, find a fuzzy incremental controller that balances the rod for the inverted pendulum (*Hint*: Initially designing a nonfuzzy incremental PD, then converting to fuzzy may help).

4.10 Consider the robotic link of Section 1.4 with mathematical model

$$\ddot{\psi} = -64\sin\psi - 5\dot{\psi} + 4i$$

where i is the current delivered to the motor.

Design a fuzzy controller to make the link angle $\psi(t)$ track a reference angle $r(t)$. Let the inputs to the controller be ψ, $\dot{\psi}$ and the output be the motor current $i(t)$. (a) Define three fuzzy sets Negative (N), Zero (Z), and Positive (P) on each universe. (b) Write the rule base for your controller. (c) Simulate the closed-loop system for an initial condition of $[\psi, \dot{\psi}] = \left[\frac{\pi}{2}, 0\right]$ and a reference signal of

$$r(t) = \frac{\pi}{2} + \sin \pi t$$

Use a fourth order Runge–Kutta integration routine with a step size of 0.001 s.

CHAPTER 5

MODELING AND CONTROL METHODS USEFUL FOR FUZZY CONTROL

Basic fuzzy control, unlike most control methods, is not based on a mathematical model of the process being controlled. This is one of the strengths of fuzzy control. However, more advanced fuzzy control methods, such as some types of parallel distributed compensation and fuzzy adaptive control, as well as fuzzy system identification, do require at least an assumption of some particular structure of the model. Some methods assume continuous-time linear or nonlinear state-space model structures while others assume discrete-time state space or input–output difference equation model structures.

We emphasize that a particular dynamic system can be modeled with any of these model structures, as will be demonstrated below. The particular structure used depends on the one required by the control or identification method. Therefore, we give a brief summary of several well-known model structures for dynamic systems.

5.1 CONTINUOUS-TIME MODEL FORMS

The four most common methods for describing continuous-time dynamic systems are the transfer function (for time-invariant linear systems), the impulse response (for time-varying or time-invariant linear systems), the input–output ordinary differential equation (for any type of continuous-time system), and the state-space description (for any type of system). Of these, only the linear or nonlinear time-invariant state-space descriptions are useful for fuzzy identification and control. Now we give brief summaries of these model structures.

5.1.1 Nonlinear Time-Invariant Continuous-Time State-Space Models

Let $x(t) = [x_1, x_2, \dots, x_n]$ be the vector of states of a time-invariant continuous-time nth order single-input, single-output nonlinear system with input $u(t)$ and output $y(t)$. A very general form for the system model is

Fuzzy Control and Identification, By John H. Lilly

$$\dot{x} = F(x, u) \tag{5.1a}$$
$$y = H(x, u) \tag{5.1b}$$

where $F(x, u)$ and $H(x, u)$ are continuously differentiable functions of their arguments.

If (5.1) can be put in the form

$$\dot{x} = f(x) + g(x)u \tag{5.2a}$$
$$y = h(x) \tag{5.2b}$$

it is said to be *feedback linearizable* (a mathematically rigorous definition of feedback linearizability is more involved [1]. This means that it can be linearized by an appropriately designed feedback law. Under certain assumptions, by differentiating the output it is possible to transform (5.2) into the so-called *companion form* [21]:

$$y^{(m)} = \delta(x) + \eta(x)u \tag{5.3}$$

where $y^{(m)} = d^m y/dt^m$. The integer m is known as the *relative degree* of the system. If $m < n$, there can be *zero dynamics* ([1,21,22]), which are assumed stable in this book.

Models of the form (5.3) are used in indirect adaptive fuzzy control algorithms. In such algorithms, the system model is not known in advance; rather it is *identified* in this form using one of several fuzzy identification schemes (gradient, least squares, etc.) (see Chapter 9).

EXAMPLE 5.1

Consider the forced rigid pendulum shown in Figure 5.1. It has a massless shaft of length L with a mass M concentrated at the end, and a coefficient of friction B at the attach point. The external torque applied to the shaft at the attach point is τ. This is a version of the motor-driven robotic link of Section 1.4.

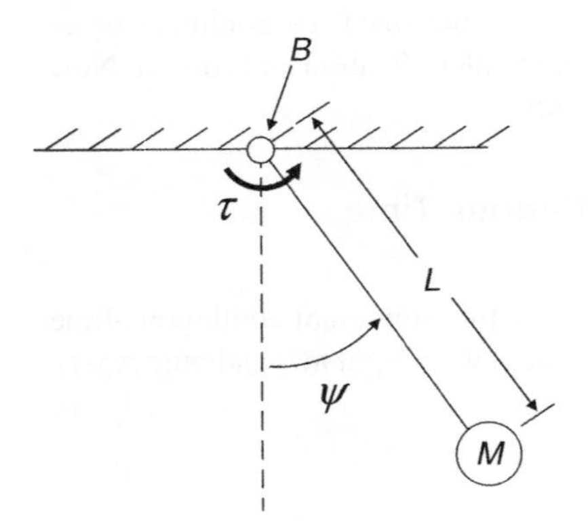

Figure 5.1. Forced rigid pendulum.

The rotational version of Newton's second law gives $\Upsilon = I\ddot{\psi}$, where ψ is the angle of the shaft from vertical, I is the moment of inertia of the pendulum about the attach point, and Υ is the sum of all torques acting on the shaft. Thus, the mathematical model of this system is

$$\tau - B\dot{\psi} - MgL\sin\psi = I\ddot{\psi}$$

where $I = ML^2$ and g is the acceleration of gravity. Let $M = 1\,\text{kg}$, $L = 1\,\text{m}$, $B = 1\,\text{kg-m/s}^2$ and $g = 9.81\,\text{m/s}^2$. If we define the states $x_1 = \psi$ and $x_2 = \dot{\psi}$, the output $y = \psi$, and the input $u = \tau$, the following state and output equations result:

$$\dot{x}_1 = x_2 \qquad (5.4a)$$

$$\dot{x}_2 = -9.8\sin x_1 - x_2 + u \qquad (5.4b)$$

$$y = x_1 \qquad (5.4c)$$

This is in the form of (5.1) with $x = [x_1 \quad x_2]^{\mathrm{T}}$,

$$F(x,u) = \begin{bmatrix} F_1 \\ F_2 \end{bmatrix} = \begin{bmatrix} x_2 \\ -9.8\sin x_1 - x_2 + u \end{bmatrix} \qquad (5.5)$$

and

$$H(x,u) = x_1$$

The right-hand sides of (5.4a) and (5.4b) can be rewritten to give

$$\begin{bmatrix} \dot{x}_1 \\ \dot{x}_2 \end{bmatrix} = \begin{bmatrix} x_2 \\ -9.8\sin x_1 - x_2 \end{bmatrix} + \begin{bmatrix} 0 \\ 1 \end{bmatrix} u \qquad (5.6a)$$

$$y = x_1 \qquad (5.6b)$$

Since this is the form of (5.2) with $f(x) = \begin{bmatrix} x_2 \\ -9.8\sin x_1 - x_2 \end{bmatrix}$, $g(x) = \begin{bmatrix} 0 \\ 1 \end{bmatrix}$, and $h(x) = x_1$, the system is feedback linearizable.

Differentiating (5.6b) once, we have $\dot{y} = \dot{x}_1 (= x_2)$. This is not in the form of (5.3) because there is no u term present. Differentiating the output once again, we have $\ddot{y} = \dot{x}_2$. Then, the model can be expressed in the form of (5.3) with relative degree $m = 2$, $\delta(x) = -9.8\sin x_1 - x_2$ and $\eta(x) = 1$:

$$\ddot{y} = -9.8\sin x_1 - x_2 + u \qquad (5.7)$$

The Matlab code to simulate the pendulum model (5.4) forced with an input torque of

$$u = \begin{cases} 2\sin(2\pi t), & t \le 2\,\text{s} \\ 0, & t > 2\,\text{s} \end{cases} \qquad (5.8)$$

is in the Appendix . The resulting pendulum angle is shown in Figure 5.2.

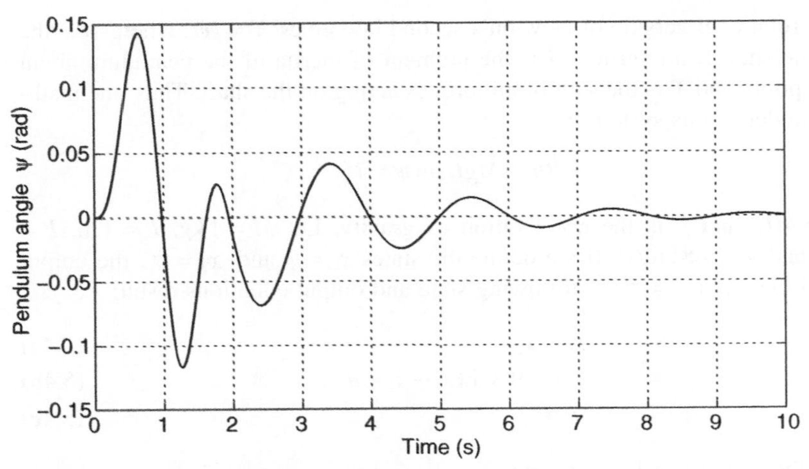

Figure 5.2. Pendulum angle, full nonlinear model (5.1).

5.1.2 Linear Time-Invariant Continuous-Time State-Space Models

Let the nonlinear system be modeled as in (5.1). Assume $x = 0$, $u = 0$ is an equilibrium point of the system [i.e., $F(0, 0) = 0$]. Then expanding $F(x, u)$ in a Taylor series about $x = 0$, $u = 0$, we have

$$F(x,u) = F(0,0) + \left.\frac{\partial F}{\partial x}\right|_{\substack{x=0\\u=0}} x + \left.\frac{\partial F}{\partial u}\right|_{\substack{x=0\\u=0}} u + \ldots$$

In this series, the higher-order terms (i.e., those involving higher than first powers of x or u or cross-products between these) can be ignored in the vicinity of the equilibrium point because these terms are vanishingly small in this region. Therefore, in the vicinity of $x = 0$, $u = 0$, the system is approximately described by the linear model

$$\dot{x}(t) = Ax(t) + bu(t) \tag{5.9a}$$

$$y(t) = cx(t) \tag{5.9b}$$

where $A = \left.\frac{\partial F}{\partial x}\right|_{\substack{x=0\\u=0}}$, $b = \left.\frac{\partial F}{\partial u}\right|_{\substack{x=0\\u=0}}$, and $c = \left.\frac{\partial H}{\partial x}\right|_{\substack{x=0\\u=0}}$

EXAMPLE 5.2

The model of the forced pendulum of Example 5.1 can be expressed as (5.1) with $F(x, u)$ as in (5.5). Then,

$$\left.\frac{\partial F}{\partial x}\right|_{\substack{x=0\\u=0}} = \left.\begin{bmatrix} \dfrac{\partial F_1}{\partial x_1} & \dfrac{\partial F_1}{\partial x_2} \\ \dfrac{\partial F_2}{\partial x_1} & \dfrac{\partial F_2}{\partial x_2} \end{bmatrix}\right|_{\substack{x=0\\u=0}} = \left.\begin{bmatrix} 0 & 1 \\ -9.8\cos x_1 & -1 \end{bmatrix}\right|_{\substack{x=0\\u=0}} = \begin{bmatrix} 0 & 1 \\ -9.8 & -1 \end{bmatrix}$$

$$\left.\frac{\partial F}{\partial u}\right|_{\substack{x=0\\u=0}} = \left.\begin{bmatrix} \dfrac{\partial F_1}{\partial u} \\ \dfrac{\partial F_2}{\partial u} \end{bmatrix}\right|_{\substack{x=0\\u=0}} = \begin{bmatrix} 0 \\ 1 \end{bmatrix}$$

This gives the model (5.9) linearized about the equilibrium point $x_1 = 0$, $x_2 = 0$, $u = 0$:

$$\dot{x}(t) = \begin{bmatrix} 0 & 1 \\ -9.8 & -1 \end{bmatrix} x(t) + \begin{bmatrix} 0 \\ 1 \end{bmatrix} u(t) \tag{5.10a}$$

$$y(t) = \begin{bmatrix} 1 & 0 \end{bmatrix} x(t) \tag{5.10b}$$

When (5.10) is simulated with the input torque of (5.8), the behavior of the linearized system is almost identical to that of the original nonlinear system (5.4). This occurs because $\psi(t)$ and $\dot{\psi}(t)$ remain close to their equilibrium values of $\psi = 0$, $\dot{\psi} = 0$, hence higher-order terms in the Taylor series are negligibly small.

5.2 MODEL FORMS FOR DISCRETE-TIME SYSTEMS

Most systems to be controlled are continuous-time systems. If such a system is to be controlled by connecting it to a digital computer, which is virtually always the case with fuzzy control, the control input must eventually be represented in discrete-time form, necessitating some type of method that approximates continuous-time signals by discrete-time signals. This could be done by calculating the continuous-time control signal from a continuous-time model in the form of (5.1), (5.3), or (5.9) and then discretizing it, or by calculating a discrete-time control signal from a discrete-time model of the system. In the former case, the approximation error is the result of discretizing a continuous-time control signal. In the latter case, the error is the result of approximating a continuous-time system by a discrete-time system.

If the system is discrete time, either inherently or through sampling, all signals are considered as existing only at discrete time instants, $k\Delta t$, where $k = 0, 1, 2, \ldots$ and Δt is the sampling interval for sampled continuous-time systems. If the system is inherently discrete, $\Delta t = 1$. For sampled continuous-time systems, the sampling interval Δt is usually supressed, that is, $x(k\Delta t)$ is written $x(k)$.

In this section, we give two forms of discrete-time models for linear systems.

5.2.1 Input–Output Difference Equation Model for Linear Discrete-Time Systems

One common method of describing the input–output behavior of a discrete-time linear system is by an input–output difference equation involving present and past inputs and outputs. This is known as an *autoregressive moving average* (ARMA) model. The general form of ARMA model used in this book is

$$y(k+1)=\alpha(q^{-1})y(k)+\beta(q^{-1})u(k) \tag{5.11}$$

with

$$\alpha(q^{-1})=a_1+a_2q^{-1}+\cdots+a_nq^{-(n-1)} \tag{5.12a}$$
$$\beta(q^{-1})=b_0+b_1q^{-1}+\cdots+b_mq^{-m} \tag{5.12b}$$

where k is an integer representing time, $u(k)$ is the system input at time k, $y(k)$ is the system output at time k, a_i, $i = 1, 2, \ldots, n$ and b_j, $j = 0, 1, \ldots, m$ are constants, and q^{-1} is the backward-shift operator defined by

$$q^{-1}y(k)=y(k-1) \tag{5.13}$$

By incrementing the time index in (5.11) by $n-1$, the following model results:

$$\begin{aligned}&y(k+n)-a_1y(k+n-1)-a_2y(k+n-2)-\cdots-a_ny(k)\\&=b_0u(k+n-1)+b_1u(k+n-2)+\cdots+b_mu(k+n-m-1)\end{aligned} \tag{5.14}$$

Implementing the forward-shift operator q in (5.13) results in

$$\begin{aligned}&q^ny(k)-a_1q^{n-1}y(k)-a_2q^{n-2}y(k)-\cdots-a_ny(k)\\&=b_0q^{n-1}u(k)+b_1q^{n-2}u(k)+\cdots+b_mq^{n-m-1}u(k)\end{aligned}$$

from which we can write the system pulse transfer function:

$$\frac{y(k)}{u(k)}=\frac{b_0q^{n-1}+b_1q^{n-2}+\cdots+b_mq^{n-m-1}}{q^n-a_1q^{n-1}-a_2q^{n-2}-\cdots-a_n} \tag{5.15}$$

5.2.2 Linear Time-Invariant Discrete-Time State-Space Models

If we define the states $x_1(k) = y(k)$, $x_2(k) = y(k + 1)$, $x_3(k) = y(k + 2)$, $\ldots$, $x_n(k) = y(k + n - 1)$ in (5.14), the following state equations result

$$x(k+1)=Ax(k)+bu(k) \tag{5.16a}$$
$$y(k)=cx(k) \tag{5.16b}$$

where $x(k) = [x_1(k)\ x_2(k)\ \ldots\ x_n(k)]^T$ and

$$A=\begin{bmatrix} 0 & 1 & 0 & \cdots & 0 \\ 0 & 0 & 1 & \cdots & 0 \\ 0 & 0 & 0 & \ddots & \vdots \\ \vdots & \vdots & \vdots & \cdots & 1 \\ a_n & a_{n-1} & a_{n-2} & \cdots & a_1 \end{bmatrix}, b=\begin{bmatrix} 0 \\ 0 \\ 0 \\ \vdots \\ 1 \end{bmatrix}, c^T=\begin{bmatrix} 0 \\ \vdots \\ b_m \\ \vdots \\ b_0 \end{bmatrix}$$

The above gives a method for converting from input–output difference equation form to linear discrete-time state-space form. It is also possible to convert to linear discrete-time state-space form directly from linear continuous-time state space form (5.9). To do this, discretize time and define $x(k) = x(k\Delta t)$, where Δt is the sample time and k is an integer. Approximate $\dot{x}$ as

$$\dot{x} \approx \frac{x(k+1)-x(k)}{\Delta t}$$

Then (5.9) becomes

$$\frac{x(k+1)-x(k)}{\Delta t} \approx Ax(k)+bu(k)$$
$$y(k)=cx(k)$$

which yields

$$x(k+1)=(I+A\Delta t)x(t)+b\Delta tu(k)$$
$$y(k)=cx(k)$$

EXAMPLE 5.3

Consider the linearized continuous-time model of the forced pendulum (5.10). If the sampling time Δt is 0.01 s, the discrete-time linearized model of the forced pendulum is

$$x(k+1)=\begin{bmatrix} 1 & 0.01 \\ -0.098 & 0.99 \end{bmatrix}x(k)+\begin{bmatrix} 0 \\ 0.01 \end{bmatrix}u(k) \tag{5.17a}$$

$$y(k)=[1 \quad 0]x(k) \tag{5.17b}$$

□

The Matlab code to simulate the pendulum model (5.17) forced with the input torque of (5.8) is given in the Appendix. The behavior of the linearized discrete-time system is nearly identical to that of the original nonlinear system. This is because the sampling time $\Delta t = 0.01$ s is small enough that all signals are essentially constant over one sampling interval.

EXAMPLE 5.4

Applying the forward shift operator (5.13) to (5.17) yields

$$qx(k)=\begin{bmatrix}1 & 0.01\\ -0.098 & 0.99\end{bmatrix}x(k)+\begin{bmatrix}0\\ 0.01\end{bmatrix}u(k)$$

$$y(k)=[1 \quad 0]x(k)$$

Solving for $x(k)$ in the first equation and substituting this into the second equation yields

$$y(k)=[1 \quad 0]\left(qI-\begin{bmatrix}1 & 0.01\\ -0.098 & 0.99\end{bmatrix}\right)^{-1}\begin{bmatrix}0\\ 0.01\end{bmatrix}u(k)$$

Therefore, the pulse transfer function of the forced pendulum is

$$\frac{y(k)}{u(k)}=\frac{0.0001}{q^2-1.99q+0.991}$$

which gives rise to the linear input–output difference equation in the form of (5.11) describing the forced pendulum:

$$y(k+1)=1.99y(k)-0.991y(k-1)+0.0001u(k-1) \tag{5.18}$$

The Matlab code to simulate the pendulum model (5.18)) forced with the input torque of (5.8) is given in the Appendix . The behavior of the system (5.18) is essentially identical to that of the original nonlinear system.

It should be emphasized that the models (5.4), (5.7), (5.10), (5.17), and (5.18) all describe the forced pendulum under certain conditions. Each is inaccurate in its own way. The original nonlinear model (5.4) is perhaps the least inaccurate, but even it is still only an approximation of the true physical pendulum. This is due to several reasons: (1) Equation (5.4) assumes the pendulum shaft is massless and that the mass M is concentrated at one point at the end of the shaft. This is not true for an actual pendulum. (2) Equation (5.4) assumes that the shaft is absolutely rigid. In an actual pendulum, the shaft is flexible to some extent. (3) Equation (5.4) assumes that the quantities M, B, and L are precisely known. In reality, they are only approximately known, and B may change over time. (4) The input torque τ is not known with precision. If the difficulties (1–4) listed above were not present, that is, if an absolutely accurate model were used to describe the pendulum, it would be impossibly complex. This gives support for using the various models above, since all mathematical models are inaccurate to some extent.

5.3 SOME CONVENTIONAL CONTROL METHODS USEFUL IN FUZZY CONTROL

The advent of Takagi–Sugeno (T–S) fuzzy systems ([8,9,11,12,23–25]) has enabled the use of some well-known conventional control methods for control of nonlinear systems with unknown models. Below we give the methods that will be used in later chapters in conjunction with T–S fuzzy systems.

5.3.1 Pole Placement Control [26]

If the system model is of the linear time-invariant state-space form [(5.9) for continuous-time systems or (5.16) for discrete-time systems], properties like stability, rise time, settling time, overshoot, and so on, of the system depend on the system's poles, which are the eigenvalues of the dynamic matrix A. If the eigenvalues are not as desired, they can be moved to desired locations via full state feedback if the system is controllable, that is, if

$$\mathcal{C} = \begin{bmatrix} b & Ab & A^2b & \cdots & A^{n-1}b \end{bmatrix} \tag{5.19}$$

is nonsingular. The matrix $\mathcal{C}$ is called the *controllability matrix*.

Let the characteristic polynomial of A be

$$|\lambda I - A| = \lambda^n + d_{n-1}\lambda^{n-1} + \cdots + d_1\lambda + d_0 \tag{5.20}$$

The eigenvalues of A are the roots of this polynomial. The full-state feedback control law is

$$u(t) = -kx(t) \tag{5.21}$$

When this input is applied to system (5.9), the resulting closed-loop system is

$$\dot{x}(t) = (A - bk)x(t) \tag{5.22}$$

Let the desired closed-loop eigenvalues be $\mu_1, \mu_2, \ldots, \mu_n$. These are the desired eigenvalues of $A - bk$. Then the desired characteristic polynomial of $A - bk$ is

$$(\lambda - \mu_1)(\lambda - \mu_2)\cdots(\lambda - \mu_n) = \lambda^n + \delta_{n-1}\lambda^{n-1} + \cdots + \delta_1\lambda + \delta_0 \tag{5.23}$$

Define

$$W = \begin{bmatrix} d_1 & d_2 & \cdots & d_{n-1} & 1 \\ d_2 & d_3 & \cdots & 1 & 0 \\ \vdots & \vdots & & \vdots & \vdots \\ d_{n-1} & 1 & \cdots & 0 & 0 \\ 1 & 0 & \cdots & 0 & 0 \end{bmatrix} \tag{5.24}$$

Then the full-state feedback control law that places the closed-loop poles at $\mu_1, \mu_2, \ldots, \mu_n$ is given by (5.21), where

$$k = [\delta_0 - d_0, \quad \delta_1 - d_1, \quad \cdots, \quad \delta_{n-2} - d_{n-2}, \quad \delta_{n-1} - d_{n-1}]T^{-1} \tag{5.25}$$

with

$$T = \mathcal{C}W \tag{5.26}$$

Note that this method applies for discrete-time systems like (5.16) as well.

EXAMPLE 5.5

Consider the linear continuous-time time-invariant system given by (5.9) with

$$A = \begin{bmatrix} -1 & 0 & 1 & 0 \\ 0 & 2 & 1 & -1 \\ 0 & 0 & 0 & 1 \\ 0 & -1 & 1 & 0 \end{bmatrix}, b = \begin{bmatrix} 1 \\ 0 \\ 0 \\ -1 \end{bmatrix}$$

Design a linear state feedback control law $u = -kx$ such that the poles of the closed-loop system are -1, $-1 + j$, $-1 - j$, -2.

Solution: The characteristic polynomial of A is

$$|\lambda I - A| = \lambda^4 - \lambda^3 - 4\lambda^2 + \lambda + 3$$

This can also be found using the Matlab command *poly*. Then, we have $d_0 = 3$, $d_1 = 1$, $d_2 = -4$, $d_3 = -1$ and

$$W = \begin{bmatrix} 1 & -4 & -1 & 1 \\ -4 & -1 & 1 & 0 \\ -1 & 1 & 0 & 0 \\ 1 & 0 & 0 & 0 \end{bmatrix}$$

The controllability matrix of the system is

$$C = \begin{bmatrix} b & Ab & A^2b & A^3b \end{bmatrix} = \begin{bmatrix} 1 & -1 & 0 & 0 \\ 0 & 1 & 1 & 4 \\ 0 & -1 & 0 & -2 \\ -1 & 0 & -2 & -1 \end{bmatrix}$$

Therefore,

$$T = CW = \begin{bmatrix} 5 & -3 & -2 & 1 \\ -1 & 0 & 1 & 0 \\ 2 & 1 & -1 & 0 \\ 0 & 2 & 1 & -1 \end{bmatrix}$$

The desired closed-loop characteristic polynomial is

$$(\lambda + 1)(\lambda + 1 - j)(\lambda + 1 + j)(\lambda + 2) = \lambda^4 + 5\lambda^3 + 10\lambda^2 + 10\lambda + 4$$

This can also be found using the Matlab command *poly*. Then we have $\delta_0 = 4$, $\delta_1 = 10$, $\delta_2 = 10$, $\delta_3 = 5$. Then, from (5.25), the feedback matrix k is

$$k = [1 \quad 9 \quad 14 \quad 6]T^{-1} = [0 \quad 41 \quad 21 \quad -6]$$

and the feedback control law is

$$u = -kx = -41x_2 - 21x_3 + 6x_4 \tag{5.27}$$

To check, verify that the eigenvalues of $A - bk$ are $-1, -1 + j, -1 - j, -2$ as desired.

□

If the number of states is small (say $n \le 3$), it may be easier to calculate k by simply calculating the characteristic polynomial of $A - bk$ with variable k and equating this to the desired closed-loop characteristic polynomial.

EXAMPLE 5.6

Consider the discrete-time linear time-invariant system given by (5.16) with

$$A = \begin{bmatrix} 0.1 & -1.5 \\ 1.2 & 0 \end{bmatrix}, b = \begin{bmatrix} 0 \\ 1 \end{bmatrix}$$

Design a linear state feedback control law $u = -kx$ such that the poles of the closed-loop system are $0.8 \pm j0.2$.

Solution: Since this is only a two-state system, we can find k by calculating the characteristic polynomial of $A - bk$ with variable k and equating this to the desired closed-loop characteristic polynomial. The desired characteristic polynomial is

$$(\lambda - 0.8 - j0.2)(\lambda - 0.8 + j0.2) = \lambda^2 - 1.6\lambda + 0.68 \tag{5.28}$$

If $k = [k_1 \quad k_2]$, the closed-loop characteristic polynomial is

$$\begin{aligned} |\lambda I - A + bk| &= \begin{vmatrix} \lambda - 0.1 & 1.5 \\ -1.2 + k_1 & \lambda + k_2 \end{vmatrix} \\ &= \lambda^2 + (k_2 - 0.1)\lambda + (-1.5k_1 - 0.1k_2 + 1.8) \end{aligned} \tag{5.29}$$

Equating coefficients in (5.28) and (5.29) yields $k_2 - 0.1 = -1.6$ and $-1.5k_1 - 0.1k_2 + 1.8 = 0.68$, from which $k_1 = 0.8467$ and $k_2 = -1.5$. Thus the feedback control law that places the closed-loop eigenvalues at $0.8 \pm j0.2$ is

$$u = -kx = -0.8467x_1 + 1.5x_2 \tag{5.30}$$

5.3.2 Tracking Control [27]

Assume that an nth-order SISO linear discrete-time system has unit delay (a similar development applies for systems with delays >1). Then it has an input–output ARMA model as in (5.11) and (5.12).

The one-step-ahead tracking control problem is to find a control $u(k)$ that brings the output at time $k + 1(y(k + 1))$ to some desired bounded value $r(k + 1)$ and is such that $y(k) = r(k)$ thereafter. It is easy to verify that the control that accomplishes this satisfies

$$\beta(q^{-1})u(k) = r(k+1) - \alpha(q^{-1})y(k) \tag{5.31}$$

To see this, substitute (5.31) into (5.11). The resulting closed-loop system has bounded inputs and outputs provided that the poles of the inverse model (i.e., the zeros of the polynomial $q^m\beta(q^{-1})$ lie inside the unit circle. This design procedure applies to n-step-ahead tracking problems with obvious modifications.

EXAMPLE 5.7

Consider a linear time-invariant discrete-time system described by the second-order input–output difference equation:

$$y(k+1) = 1.5y(k) - 0.4(k-1) + u(k) + 0.6u(k-1) \tag{5.32}$$

Find a control $u(k)$ that brings the output at time $k + 1$ to the desired reference signal $r(k) = 0.1\sin(0.1\pi k)$.

Solution: The system can be rewritten as (5.11) with $\alpha(q^{-1}) = 1.5 - 0.4q^{-1}$ and $\beta(q^{-1}) = 1 + 0.6q^{-1}$. Note that this system is unstable because the polynomial $q^2[1 - q^{-1}\alpha(q^{-1})] = q^2 - 1.5q + 0.4$ has roots outside the unit circle. Also, we are assured the tracking objective can be accomplished with a bounded control because the poles of the inverse model [i.e., the zeros of the polynomial $q\beta(q^{-1})$] are inside the unit circle.

The control law (5.31) yields the one-step-ahead tracking controller:

$$u(k) = -0.6u(k-1) - 1.5y(k) + 0.4y(k-1) + 0.1\sin(0.1\pi(k+1)) \tag{5.33}$$

5.3.3 Model Reference Control [27]

The objective of model reference control is to find a plant input such that the output of the controlled system tracks that of a reference model. The following development is a generalization of the tracking controller above. Let the plant be described by the input–output ARMA model (5.11) and (5.12). A similar development applies for systems with delays greater than 1.

Assume, without loss of generality, that the desired output $y^*(k)$ satisfies the following reference model:

$$E(q^{-1})y^*(k) = q^{-1}gH(q^{-1})r(k) \tag{5.34}$$

where g is a constant gain, $r(k)$ is an arbitrary bounded input, and

$$H(q^{-1}) = 1 + h_1q^{-1} + \cdots + h_lq^{-l} \tag{5.35a}$$
$$E(q^{-1}) = 1 + e_1q^{-1} + \cdots + e_lq^{-l} \tag{5.35b}$$

The associated pulse transfer function of the reference model is

$$G(q^{-1}) = \frac{y^*(k)}{r(k)} = \frac{q^{-1} gH(q^{-1})}{E(q^{-1})} \tag{5.36}$$

The model reference control problem is to find a control $u(k)$ that brings the plant output $y(k)$ to the output $y^*(k)$ of the reference model (5.34) excited by a known bounded signal $r(k)$.

The control $u(k)$ that accomplishes $y(k) = y^*(k)$ satisfies

$$\gamma(q^{-1})y(k) + \beta(q^{-1})u(k) = gH(q^{-1})r(k) \tag{5.37}$$

where $\gamma(q^{-1})$ is the unique polynomial of order $n - 1$ satisfying

$$E(q^{-1}) = 1 + q^{-1}[\gamma(q^{-1}) - \alpha(q^{-1})] \tag{5.38}$$

To see this, realize that from (5.38) we have

$$\gamma(q^{-1}) = q[E(q^{-1}) - 1] + \alpha(q^{-1})$$

Substituting this in (5.37) yields

$$qE(q^{-1})y(k) - y(k+1) + \alpha(q^{-1})y(k) + \beta(q^{-1})u(k) = gH(q^{-1})r(k)$$

Since the plant model is given by (5.11), this reduces to (5.34).

EXAMPLE 5.8

Consider a system described by the ARMA model:

$$y(k+1) = 0.4y(k) - 1.8y(k-1) + 1.2u(k) + u(k-1)$$

Let the desired reference signal obey (5.36) with $g = 2$ and

$$H(q^{-1}) = 1 + 0.3q^{-1}$$
$$E(q^{-1}) = 1 + 0.5q^{-1} + 0.5q^{-2}$$

Find a control $u(k)$ that brings the plant output to the output of the reference model (5.36) excited by a known bounded signal $r(k)$.

Solution: This system is in the form of (5.11) with $\alpha(q^{-1}) = 0.4 - 1.8q^{-1}$ and $\beta(q^{-1}) = 1.2 + q^{-1}$. Note that the system is unstable because the polynomial $q^2[1 - q^{-1}\alpha(q^{-1})] = q^2 - 0.4q + 1.8$ has roots outside the unit circle. Also, we are assured that the tracking objective can be accomplished with a bounded control because the poles of the inverse model (i.e., the zeros of the polynomial $q\beta(q^{-1})$) are inside the unit circle.

The model pulse transfer function is

$$G(q^{-1}) = \frac{2q^{-1} + 0.6q^{-2}}{1 + 0.5q^{-1} + 0.5q^{-2}}$$

The identity (5.38) is given by

$$\left(1+0.5q^{-1}+0.5q^{-2}\right)=1-0.4q^{-1}+1.8q^{-2}+q^{-1}\left(\gamma_0+\gamma_1 q^{-1}\right)$$

Therefore, we have

$$\gamma\left(q^{-1}\right)=\gamma_0+\gamma_1 q^{-1}=0.9-1.3q^{-1}$$

and control $u(k)$ that accomplishes model following obeys (5.37):

$$\left(0.9-1.3q^{-1}\right)y(k)+\left(1.2+q^{-1}\right)u(k)=2\left(1+0.3q^{-1}\right)r(k)$$

This yields the model reference control law:

$$u(k)=\frac{1}{1.2}\left[-u(k-1)-0.9y(k)+1.3y(k-1)+2r(k)+0.6r(k-1)\right]$$

□

5.3.4 Feedback Linearization

Let the plant be modeled as in (5.3). Assume further that $\eta(x(t)) \geq \eta_0 > 0$, so that the system is controllable for all t. Let the control law be

$$u(t)=\frac{1}{\eta(x(t))}\left(-\delta(x(t))+v(t)\right) \tag{5.39}$$

where $v(t)$ is an arbitrary function of t chosen to produce a stable closed-loop system with desired characteristics.

Then, combining (5.3) and (5.39), the closed-loop system is given by:

$$y^{(m)}=\delta(x(t))+\eta(x(t))\left[\frac{1}{\eta(x(t))}\left(-\delta(x(t))+v(t)\right)\right]$$

or

$$y^{(m)}=v(t) \tag{5.40}$$

This is a linear system with input $v(t)$. Thus the nonlinear system has been linearized by the feedback (5.39).

EXAMPLE 5.9

For the system (5.3), let the input be given by (5.39) with

$$v(t)=-a_1 y^{(m-1)}-a_2 y^{(m-2)}-\cdots-a_m y+b_1 r^{(m-1)}+b_2 r^{(m-2)}+\cdots+b_m r \tag{5.41}$$

where $a_1 \ldots a_m$ and $b_1 \ldots b_m$ are arbitrary constants chosen by the designer and $r(t)$ is an external input. Then the closed-loop system is linear with transfer function:

$$\frac{Y(s)}{R(s)}=\frac{b_1 s^{m-1}+b_2 s^{m-2}+\cdots+b_m}{s^m+a_1 s^{m-1}+a_2 s^{m-2}+\cdots+a_m}$$

Thus by choice of $a_1 \ldots a_m$ and $b_1 \ldots b_m$, any desired closed-loop performance (overshoot, rise time, steady-state response, etc.) may be obtained.

EXAMPLE 5.10

Consider the forced pendulum (5.4). Find a feedback control law so that the closed-loop system is linear with unity zero-frequency gain and poles at $s=-1 \pm j$.

Solution: The desired transfer function (unity zero-frequency gain, poles at $s=-1 \pm j$) is

$$\frac{\psi(s)}{\tau(s)}=\frac{2}{s^2+2s+2}$$

Considering the model (5.7), which is equivalent to (5.4), we have $\delta(\psi,\dot{\psi})=-9.8\sin\psi-\dot{\psi}$ and $\eta(\psi,\dot{\psi})=1$.

Then by (5.39) and (5.41) the necessary control law is

$$\tau=9.8\sin\psi+\dot{\psi}-2\dot{\psi}-2\psi+2r=9.8\sin\psi-2\psi-\dot{\psi}+2r \tag{5.42}$$

where $r(t)$ is the external input to the system. The block diagram of the closed-loop system is shown in Figure 5.3.

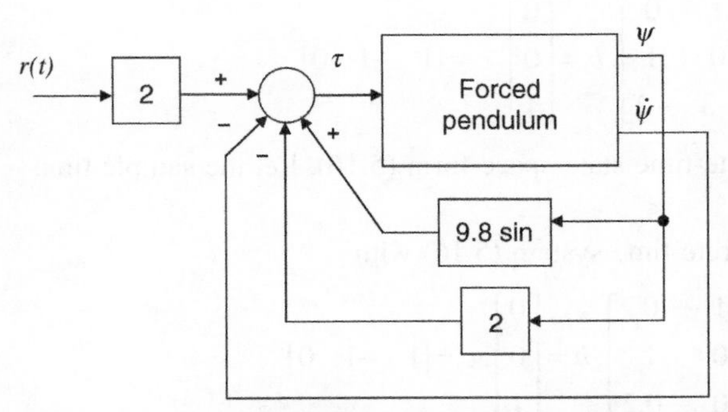

Figure 5.3. Block diagram of forced pendulum (5.4) with feedback linearizing control (5.42).

5.4 SUMMARY

This chapter contains model forms and control methodologies that are useful for fuzzy control. All of these will be used in various fuzzy control strategies implemented with T–S fuzzy systems in the following chapters. The presentation is brief, that is, some familiarity with controls is assumed. However, the theory has been kept to a minimum, with the assumption that if more detail is desired, the reader can consult one of the references given in the chapter.

The model forms considered for continuous-time systems are the **continuous-time nonlinear state equations (5.1), continuous-time linear state equations (5.9)**, and the **feedback-linearization form (5.3)**. The model forms considered for discrete-time systems are the **discrete-time linear state equations (5.16)** and **input–output difference equation, or ARMA model (5.11)**.

Continuous-time nonlinear state equations are not used directly for fuzzy control, but are used to derive models in feedback-linearizeable form. Feedback-linearizable model forms are used in indirect adaptive control schemes for adaptive model following (Section 5.3.4). Continuous- and discrete-time linear state equation model forms are used in parallel distributed compensators for pole placement (Section 5.3.1). Input–output difference equation models are used in tracking and model reference controllers (Sections 5.3.2 and 5.3.3).

EXERCISES

5.1 Put the inverted pendulum of Example 4.1 in feedback linearizable form (5.3). Consider the output to be the rod angle ψ.

5.2 Put the inverted pendulum of Example 4.1 in linearized continuous-time state-space form (5.9).

5.3 Consider the linear continuous-time state-space system (5.9) with

$$A = \begin{bmatrix} 0 & 1 & 0 \\ 0 & 0 & 1 \\ -1 & -3 & -2 \end{bmatrix}, b = \begin{bmatrix} 0 \\ 0 \\ 1 \end{bmatrix}, c = [1 \quad -1 \quad 0]$$

Put the system in discrete-time state-space form (5.16). Let the sample time be $\Delta t = 0.01$ s.

5.4 Consider the linear discrete-time system (5.16) with

$$A = \begin{bmatrix} 0 & 1 & 0 \\ 0 & 0 & 1 \\ -0.1 & 0 & -0.2 \end{bmatrix}, b = \begin{bmatrix} 0 \\ 0 \\ 1 \end{bmatrix}, c = [1 \quad -1 \quad 0]$$

Put the system in input–output difference equation form (5.11).

5.5 Consider a system described by the input–output difference equation

$$y(k+1) = 0.3y(k) - 0.6y(k-1) + 1.5u(k) + u(k-1)$$

Put the system in discrete-time state-space form (5.16).

5.6 Consider the system described by the linear continuous-time state-space equations given in Problem 5.3. Find a linear state feedback law (5.21) to place the closed-loop poles at -1, $-1 + j$, and $-1 - j$. Use the method of Example 5.5.

5.7 Repeat Problem 6 using the method of Example 5.6.

5.8 Consider a system described by the input–output difference equation

$$y(k+1)=0.4y(k)-1.8y(k-1)+1.2u(k)+u(k-1)$$

Design a one-step-ahead tracking controller to make the system's output track the reference signal $r(k) = 0.1(1 - \cos(0.3k))$.

5.9 For the system of Problem 5.8, design a model reference controller to make the system's output track the output of the model whose transfer function is

$$G(q^{-1})=\frac{q^{-1}+0.2q^{-2}}{1+0.6q^{-1}+0.7q^{-2}}$$

with the input $r(k) = 0.1(1 - \cos(0.3k))$.

5.10 Consider the system in feedback-linearizeable form

$$\ddot{y} = x_1^2 + x_2x_3 + (1+\sin^2 x_2)u$$

Design a feedback control law for this system so that the closed-loop system is linear with transfer function

$$\frac{Y(s)}{R(s)}=\frac{s^2+1}{s^3+2s^2+2s+1}$$

where $r(t)$ is an external input.

CHAPTER 6

TAKAGI–SUGENO FUZZY SYSTEMS

Takagi–Sugeno (T–S) fuzzy systems are more general than the Mamdani fuzzy systems discussed in Chapters 3 and 4. In fact, it can be shown that Mamdani systems are special cases of T–S fuzzy systems. The T–S systems are important because they enable a kind of control called *parallel distributed control*, they facilitate fuzzy identification of dynamic systems and adaptive fuzzy control, and they enable stability proofs for certain closed-loop systems involving fuzzy controllers. Their drawback is that they are less intuitive than Mamdani systems.

In T–S systems, the consequents of the rules do not involve fuzzy sets as do Mamdani systems, but instead are mathematical expressions. The mathematical expressions can be any linear functions of any variables. In this book, we consider only consequents that are either memoryless affine functions of the fuzzy system's inputs, or one of the linear dynamic system model forms discussed in Sections 5.1 or 5.2. In the former case, the T–S fuzzy system performs an interpolation between memoryless functions. In the latter case, the T–S system performs an interpolation between dynamic systems. The latter case is useful for fuzzy identification and control.

6.1 TAKAGI–SUGENO FUZZY SYSTEMS AS INTERPOLATORS BETWEEN MEMORYLESS FUNCTIONS

Consider a T–S fuzzy system with R rules of the form:

$$R_i \qquad \text{If } x_1 \text{ is } P_1^K \text{ and } x_2 \text{ is } P_2^L \text{ and} \cdots \text{and } x_n \text{ is } P_n^M, \text{then } q^i = f_i(\bullet) \tag{6.1}$$

The input universes $\mathcal{X}_1, \mathcal{X}_2, \ldots, \mathcal{X}_n$ are as before, with various fuzzy sets defined on them, but now the consequents are memoryless functions $f_i(\bullet)$. The crisp output of this system is

$$y^{\text{crisp}} = \frac{\sum_{i=1}^{R} q^i \mu_i(\underline{x})}{\sum_{i=1}^{R} \mu_i(\underline{x})} \tag{6.2}$$

Fuzzy Control and Identification, By John H. Lilly

where $\mu_i(\underline{x})$ is the premise membership value of Rule i. The crisp output can be expressed as

$$y^{\text{crisp}} = q^1\xi_1(t) + q^2\xi_2(t) + \cdots + q^R\xi_R(t) \tag{6.3}$$

where

$$\xi_i(t) = \frac{\mu_i(x(t))}{\sum_{j=1}^{R}\mu_j(x(t))} \quad i = 1, \ldots, R \tag{6.4}$$

are called *fuzzy basis functions* [28].

In this section, we assume the functions $f_i(\bullet)$ are affine functions of the inputs:

$$f_i(\bullet) = a_0^i + a_1^i x_1 + a_2^i x_2 + \cdots + a_n^i x_n \tag{6.5}$$

where the coefficients a_j^i are constants. The output y^{crisp} is a nonlinear function of the inputs $x_1, \ldots, x_n$.

EXAMPLE 6.1

Consider a two-input T–S fuzzy system with rules

1. If x_1 is P_1^1 and x_2 is P_2^1, then $q^1 = 1 + x_1 + x_2$.
2. If x_1 is P_1^1 and x_2 is P_2^2, then $q^2 = 2x_1 + x_2$.
3. If x_1 is P_1^2 and x_2 is P_2^1, then $q^3 = -1 + x_1 - 2x_2$.
4. If x_1 is P_1^2 and x_2 is P_2^2, then $q^4 = -2 - x_1 + 0.5x_2$.

Inside the effective universe of discourse the fuzzy sets P_1^1, P_1^2, P_2^1, and P_2^2 are characterized by the following membership functions:

$$\mu_1^1(x_1) = \exp\left[-\frac{1}{2}\left(\frac{x_1+1}{0.8493}\right)^2\right]$$

$$\mu_1^2(x_1) = \exp\left[-\frac{1}{2}\left(\frac{x_1-1}{0.8493}\right)^2\right]$$

$$\mu_2^1(x_2) = \exp\left[-\frac{1}{2}\left(\frac{x_2+1}{0.8493}\right)^2\right]$$

$$\mu_2^2(x_2) = \exp\left[-\frac{1}{2}\left(\frac{x_2-1}{0.8493}\right)^2\right]$$

These are shown in Figures 6.1 and 6.2.

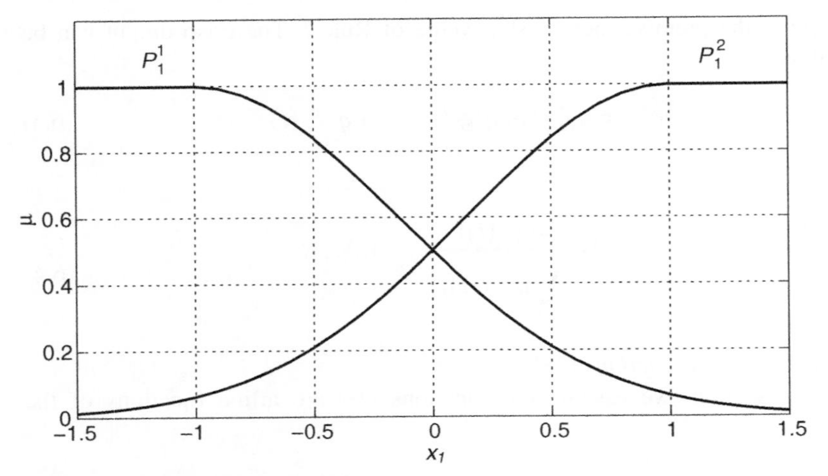

Figure 6.1. Memberships for x_1 input.

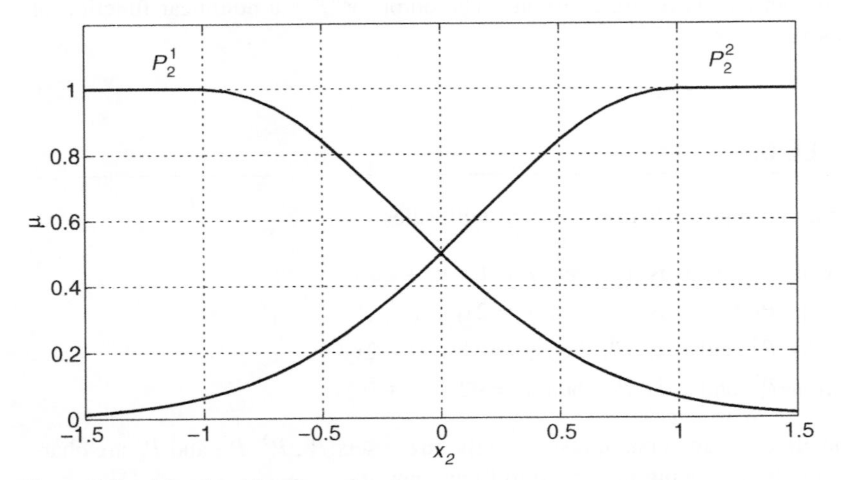

Figure 6.2. Memberships for x_2 input.

Given the crisp input (x_1, x_2), the crisp output of this system is [see (6.3), (6.4)]:

$$y^{\text{crisp}}(x_1, x_2) = \sum_{i=1}^{4} q^i(x_1, x_2)\xi_i(x_1, x_2)$$

where $\xi_i(x_1, x_2) = \mu_i(x_1, x_2) \Big/ \sum_{j=1}^{4} \mu_j(x_1, x_2)$, $i = 1, 2, 3, 4$ and, assuming *product* T-norm,

$$\mu_1(x_1, x_2) = \exp\left[-\frac{1}{2}\left(\frac{x_1+1}{0.8493}\right)^2\right]\exp\left[-\frac{1}{2}\left(\frac{x_2+1}{0.8493}\right)^2\right] \tag{6.6a}$$

$$\mu_2(x_1, x_2) = \exp\left[-\frac{1}{2}\left(\frac{x_1+1}{0.8493}\right)^2\right]\exp\left[-\frac{1}{2}\left(\frac{x_2-1}{0.8493}\right)^2\right] \quad (6.6b)$$

$$\mu_3(x_1, x_2) = \exp\left[-\frac{1}{2}\left(\frac{x_1-1}{0.8493}\right)^2\right]\exp\left[-\frac{1}{2}\left(\frac{x_2+1}{0.8493}\right)^2\right] \quad (6.6c)$$

$$\mu_4(x_1, x_2) = \exp\left[-\frac{1}{2}\left(\frac{x_1-1}{0.8493}\right)^2\right]\exp\left[-\frac{1}{2}\left(\frac{x_2-1}{0.8493}\right)^2\right] \quad (6.6d)$$

For the particular crisp input $(x_1, x_2) = (0.6, -0.3)$, we calculate $\mu_1(0.6, -0.3) = 0.1207$, $\mu_2(0.6, -0.3) = 0.0525$, $\mu_3(0.6, -0.3) = 0.6373$, and $\mu_4(0.6, -0.3) = 0.2774$. These values yield the following fuzzy basis function values:

$$\xi_1(0.6, -0.3) = 0.1207/(0.1207 + 0.0525 + 0.6373 + 0.2774) = 0.1110$$
$$\xi_2(0.6, -0.3) = 0.0525/(0.1207 + 0.0525 + 0.6373 + 0.2774) = 0.0483$$
$$\xi_3(0.6, -0.3) = 0.6373/(0.1207 + 0.0525 + 0.6373 + 0.2774) = 0.5858$$
$$\xi_4(0.6, -0.3) = 0.2774/(0.1207 + 0.0525 + 0.6373 + 0.2774) = 0.2550$$

The consequent functions are evaluated for $(x_1, x_2) = (0.6, -0.3)$ as

$$q^1(0.6, -0.3) = 1 + 0.6 - 0.3 = 1.3$$
$$q^2(0.6, -0.3) = 2(0.6) - 0.3 = 0.9$$
$$q^3(0.6, -0.3) = -1 + 0.6 - 2(-0.3) = 0.2$$
$$q^4(0.6, -0.3) = -2 - 0.6 + .5(-0.3) = -2.75$$

Therefore, the crisp output of the T–S fuzzy system corresponding to the input $(x_1, x_2) = (0.6, -0.3)$ is

$$\begin{aligned} y^{\text{crisp}}(0.6, -0.3) &= 1.3(0.1110) + 0.9(0.0483) + 0.2(0.5858) - 2.75(0.2550) \\ &= -0.3962 \end{aligned}$$

The characteristic surface of the fuzzy system is shown in Figure 6.3. This was found by evaluating the crisp output corresponding to numerous points in the domain $-2 \le x_1 < 2$, $-2 \le x_2 < 2$. The surface is an interpolation between the four affine planar functions $q^1(x_1, x_2)$, $q^2(x_1, x_2)$, $q^3(x_1, x_2)$, and $q^4(x_1, x_2)$. Small portions of these planes are shown on the corresponding edges of the characteristic surface of Figure 6.3. It is seen that on the appropriate edges of the surface, the planes match well with the surface. For instance, if x_1 and x_2 are both large negative numbers (i.e., $x_1 < -1$, $x_2 < -1$), Rule 1 would be fired with certainty, while Rules 2, 3, and 4 would be fired very little. Therefore, the crisp output of the fuzzy system would be essentially q^1.

Referring to Figure 6.3, in the area of the surface where both x_1 and x_2 are between -1 and -2, the surface is indeed essentially equal to q^1. Inside the effective universe of discourse (i.e., $-1 \le x_1 < 1$, $-1 \le x_2 < 1$), the fuzzy system interpolates between the planes.

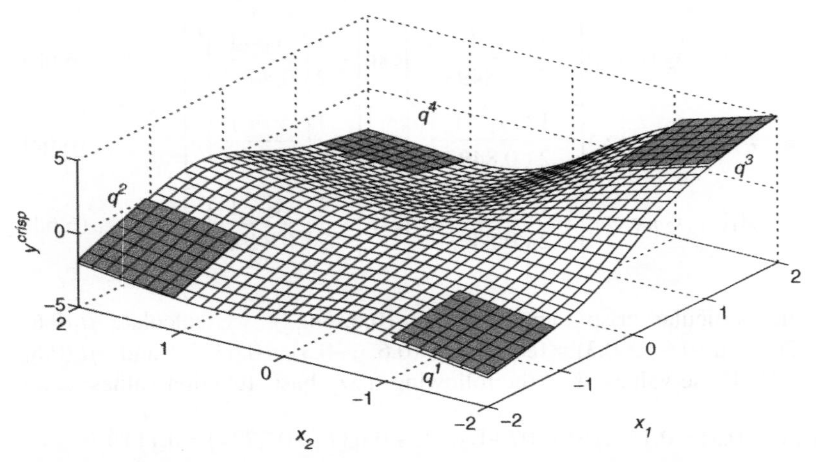

Figure 6.3. Characteristic surface of the T–S fuzzy system with individual consequents plotted at appropriate corners.

6.2 TAKAGI–SUGENO FUZZY SYSTEMS AS INTERPOLATORS BETWEEN CONTINUOUS-TIME LINEAR STATE-SPACE DYNAMIC SYSTEMS

Consider a T–S fuzzy system with R rules of the form:

R_i: If $x_1(t)$ is P_1^K and $x_2(t)$ is P_2^L and … and $x_n(t)$ is P_n^M, then

$$\dot{x}^i(t) = A_i x(t) + b_i u(t) \tag{6.7}$$

The contents of the $n \times 1$ vector $x(t) = [x_1(t), x_2(t), \ldots, x_n(t)]^T$ are the system states that vary continuously with time t. The input universes $\mathcal{X}_1, \mathcal{X}_2, \ldots, \mathcal{X}_n$ are as before, with various fuzzy sets defined on them, but now the consequents are continuous-time linear constant-coefficient dynamic systems involving the states. The quantities A_i and b_i are $n \times n$ and $n \times 1$ constant-coefficient matrices, respectively, for $i = 1, 2, \ldots, n$. This fuzzy system does not have a single crisp output, rather it produces a time-varying system described by the differential equation:

$$\dot{x}(t) = \frac{\sum_{i=1}^{R} \mu_i(x(t))(A_i x(t) + b_i u(t))}{\sum_{i=1}^{R} \mu_i(x(t))} \tag{6.8}$$

or

$$\dot{x}(t) = A(t)x(t) + b(t)u(t) \tag{6.9}$$

where $A(t) = \xi_1(t)A_1 + \xi_2(t)A_2 + \ldots + \xi_R(t)A_R$ and $b(t) = \xi_1(t)b_1 + \xi_2(t)b_2 + \ldots + \xi_R(t)b_R$ and $\xi_i(t)$, $i = 1, \ldots, R$ are the fuzzy basis functions described in (6.4). Matrices $A(t)$ and $b(t)$ contain functions of the system states, making the system (6.9) a nonlinear time-varying system.

EXAMPLE 6.2

Consider a two-input T–S fuzzy system with rules

1. If $x_1(t)$ is P_1^1 and $x_2(t)$ is P_2^1, then $\dot{x}^1(t) = A_1x(t) + b_1u(t)$.
2. If $x_1(t)$ is P_1^1 and $x_2(t)$ is P_2^2, then $\dot{x}^2(t) = A_2x(t) + b_2u(t)$.
3. If $x_1(t)$ is P_1^2 and $x_2(t)$ is P_2^1, then $\dot{x}^3(t) = A_3x(t) + b_3u(t)$.
4. If $x_1(t)$ is P_1^2 and $x_2(t)$ is P_2^2, then $\dot{x}^4(t) = A_4x(t) + b_4u(t)$.

where $A_1 = \begin{bmatrix} 0 & 1 \\ -2 & -2 \end{bmatrix}$, $b_1 = \begin{bmatrix} 0 \\ 2 \end{bmatrix}$, $A_2 = \begin{bmatrix} 0 & 1 \\ -1 & -2 \end{bmatrix}$, $b_2 = \begin{bmatrix} 2 \\ 2 \end{bmatrix}$

$$A_3 = \begin{bmatrix} 0 & 1 \\ -3 & -2 \end{bmatrix}, b_3 = \begin{bmatrix} -2 \\ 2 \end{bmatrix}, A_4 = \begin{bmatrix} 1 & 1 \\ -2 & -2 \end{bmatrix}, b_4 = \begin{bmatrix} 0 \\ -2 \end{bmatrix}$$

The fuzzy sets P_1^1, P_1^2, P_2^1, and P_2^2 are characterized by the membership functions shown in Figures 6.4 and 6.5.

For an arbitrary input (x_1, x_2) inside the effective universe of discourse (i.e., $-1 \le x_1 < 1$, $-1 \le x_2 < 1$), the premise values for the above four rules are calculated as:

$$\mu_1(t) = (-0.5x_1(t) + 0.5)(-0.25x_2(t) + 0.5) \quad (6.10a)$$
$$\mu_2(t) = (-0.5x_1(t) + 0.5)(0.25x_2(t) + 0.5) \quad (6.10b)$$
$$\mu_3(t) = (0.5x_1(t) + 0.5)(-0.25x_2(t) + 0.5) \quad (6.10c)$$
$$\mu_4(t) = (0.5x_1(t) + 0.5)(0.25x_2(t) + 0.5) \quad (6.10d)$$

where we have used the equations of the membership functions in Figures 6.4 and 6.5 inside the effective universes and *product* T-norm.

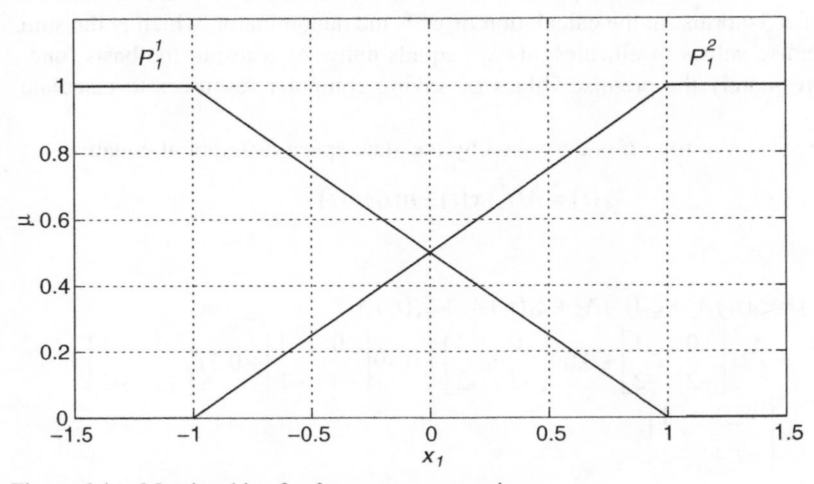

Figure 6.4. Memberships for fuzzy sets on x_1 universe.

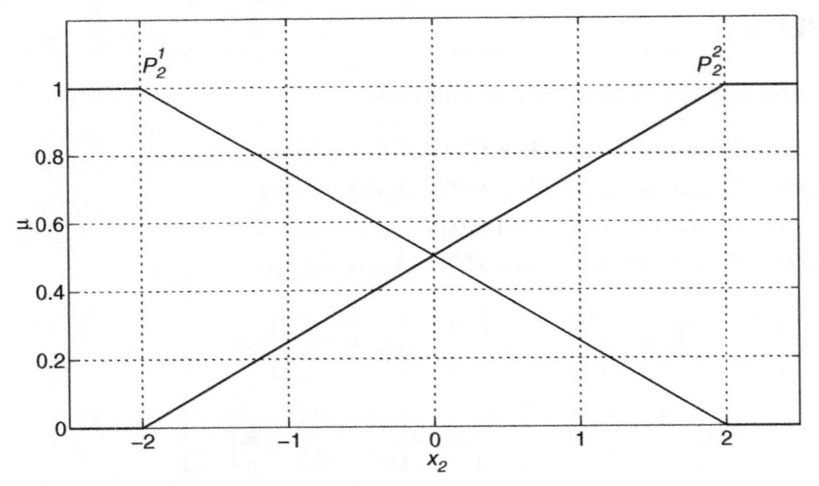

Figure 6.5. Memberships for fuzzy sets on x_2 universe.

If we consider a particular instant of time t_1 at which $x_1(t_1) = 0.4$ and $x_2(t_1) = -0.8$, we have the following premise values for the rules: $\mu_1(t_1) = 0.21$, $\mu_2(t_1) = 0.09$, $\mu_3(t_1) = 0.49$, and $\mu_4(t_1) = 0.21$. These yield the following basis function values:

$$\xi_1(t_1) = 0.21/(0.21+0.09+0.49+0.21) = 0.21$$
$$\xi_2(t_1) = 0.09/(0.21+0.09+0.49+0.21) = 0.09$$
$$\xi_3(t_1) = 0.49/(0.21+0.09+0.49+0.21) = 0.49$$
$$\xi_4(t_1) = 0.21/(0.21+0.09+0.49+0.21) = 0.21$$

Here we see the advantage of partitions of unity on the input universes combined with product T-norms: in the calculation of y^{crisp}, the denominator, which is the sum of the premise values of all rules, always equals unity. As a result, the basis functions ξ_i are merely the premise values μ_i, saving computer resources to calculate them.

The system at time t_1 is described by the state-space differential equation

$$\dot{x}(t) = A(t_1)x(t) + b(t_1)u(t)$$

where

$$\begin{aligned} A(t) &= \xi_1(t_1)A_1 + \xi_2(t_1)A_2 + \xi_3(t_1)A_3 + \xi_4(t_1)A_4 \\ &= 0.21\begin{bmatrix} 0 & 1 \\ -2 & -2 \end{bmatrix} + 0.09\begin{bmatrix} 0 & 1 \\ -1 & -2 \end{bmatrix} + 0.49\begin{bmatrix} 0 & 1 \\ -3 & -2 \end{bmatrix} + 0.21\begin{bmatrix} 1 & 1 \\ -2 & -2 \end{bmatrix} \\ &= \begin{bmatrix} 0.21 & 1 \\ -2.4 & -2 \end{bmatrix} \end{aligned}$$

and

$$b(t_1) = \xi_1(t_1)b_1 + \xi_2(t_1)b_2 + \xi_3(t_1)b_3 + \xi_4(t_1)b_4$$
$$= 0.21\begin{bmatrix}0\\2\end{bmatrix} + 0.09\begin{bmatrix}2\\2\end{bmatrix} + 0.49\begin{bmatrix}-2\\2\end{bmatrix} + 0.21\begin{bmatrix}0\\-2\end{bmatrix} = \begin{bmatrix}-0.8\\1.16\end{bmatrix}$$

To summarize, the system at time t_1 is described by

$$\begin{bmatrix}\dot{x}_1\\\dot{x}_2\end{bmatrix} = \begin{bmatrix}0.21 & 1\\-2.4 & -2\end{bmatrix}\begin{bmatrix}x_1\\x_2\end{bmatrix} + \begin{bmatrix}-0.8\\1.16\end{bmatrix}u$$

Note that this describes the system at only one time t_1. As t changes, $A(t)$ and $b(t)$ change.

A simulation of this system was carried out using fourth-order Runge–Kutta integration with a step size of $\Delta t = 0.01$ s. An input of $u(t) = 3\sin(\pi t)$ with initial condition $x_1(0) = x_2(0) = 0$ yields the state trajectories shown in Figure 6.6. The nonlinear nature of the system can be seen in this response. The Matlab simulation program producing Figure 6.6 is given in the Appendix.

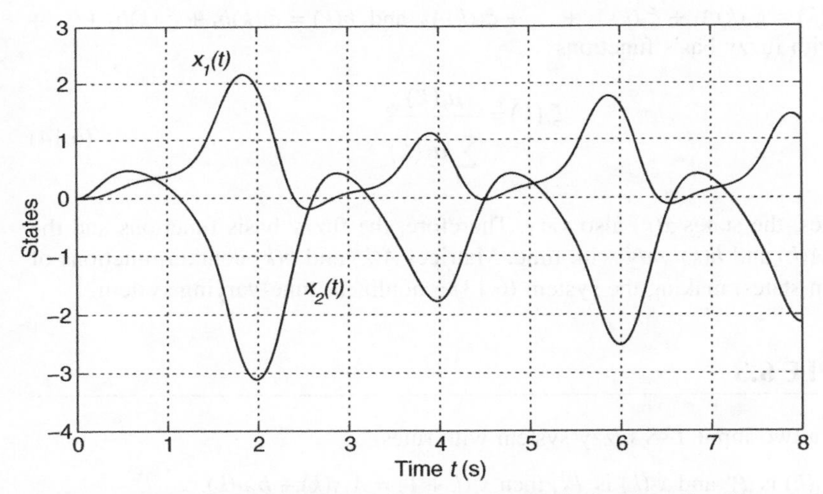

Figure 6.6 State trajectories of T–S fuzzy system that interpolates four linear continuous-time constant-coefficient state-space systems, Example 6.2.

6.3 TAKAGI–SUGENO FUZZY SYSTEMS AS INTERPOLATORS BETWEEN DISCRETE-TIME LINEAR STATE-SPACE DYNAMIC SYSTEMS

T–S fuzzy systems can be used to create time-varying discrete-time state-space dynamic systems. Consider a T–S fuzzy system with R rules of the form:

R_i: If $x_1(k)$ is P_1^K and $x_2(k)$ is P_2^L and … and $x_n(k)$ is P_n^M, then

$$x^i(k+1) = A_i x(k) + b_i x(k). \tag{6.11}$$

The contents of the $n \times 1$ vector $x(k) = [x_1(k), x_2(k), \ldots, x_n(k)]^T$ are the system states that vary with the discrete time $k \in \{0, 1, 2, 3, \ldots\}$. The input universes $\mathcal{X}_1, \mathcal{X}_2, \ldots, \mathcal{X}_n$ are as before, with various fuzzy sets defined on them, but now the consequents are discrete-time linear constant-coefficient state-space systems. This fuzzy system does not have a single crisp output, rather it produces a time-varying system described by the state-space difference equation:

$$x(k+1) = \frac{\sum_{i=1}^{R} \mu_i(k)(A_i x(k) + b_i x(k))}{\sum_{i=1}^{R} \mu_i(k)} \tag{6.12}$$

or

$$x(k+1) = A(k)x(k) + b(k)u(k) \tag{6.13}$$

where $A(k) = \xi_1(k)A_1 + \xi_2(k)A_2 + \ldots + \xi_R(k)A_R$ and $b(k) = \xi_1(k)b_1 + \xi_2(k)b_2 + \ldots + \xi_R(k)b_R$ with fuzzy basis functions

$$\xi_i(k) = \frac{\mu_i(k)}{\sum_{j=1}^{R} \mu_j(k)} \tag{6.14}$$

As k varies, the states $x(k)$ also vary. Therefore, the fuzzy basis functions and the matrices $A(k)$ and $b(k)$ vary with time. Matrices $A(k)$ and $b(k)$ contain functions of the system states, making the system (6.13) a nonlinear time-varying system.

EXAMPLE 6.3

Consider a two-input T–S fuzzy system with rules

1. If $x_1(k)$ is P_1^1 and $x_2(k)$ is P_2^1, then $x^1(k+1) = A_1x(k) + b_1u(k)$.
2. If $x_1(k)$ is P_1^1 and $x_2(k)$ is P_2^2, then $x^2(k+1) = A_2x(k) + b_2u(k)$.
3. If $x_1(k)$ is P_1^2 and $x_2(k)$ is P_2^1, then $x^3(k+1) = A_3x(k) + b_3u(k)$.
4. If $x_1(k)$ is P_1^2 and $x_2(k)$ is P_2^2, then $x^4(k+1) = A_4x(k) + b_4u(k)$.

where $A_1 = \begin{bmatrix} 0.1 & 0 \\ -0.2 & 0.2 \end{bmatrix}$, $b_1 = \begin{bmatrix} 0 \\ 1 \end{bmatrix}$, $A_2 = \begin{bmatrix} 0 & 0.1 \\ 1 & -0.2 \end{bmatrix}$, $b_2 = \begin{bmatrix} 0.1 \\ 1 \end{bmatrix}$

$$A_3 = \begin{bmatrix} -0.3 & 1 \\ 0 & 0.2 \end{bmatrix}, b_3 = \begin{bmatrix} -0.1 \\ 1 \end{bmatrix}, A_4 = \begin{bmatrix} 0.1 & 1 \\ 0.2 & 0.5 \end{bmatrix}, b_4 = \begin{bmatrix} 0 \\ -1 \end{bmatrix}$$

Let the fuzzy sets P_1^1, P_1^2, P_2^1, and P_2^2 be characterized by the Gaussian membership functions shown in Figures 6.1 and 6.2. Then for an arbitrary input (x_1, x_2) and *product* T-norm, the premise values for the above four rules are calculated as in (6.6).

If we consider a particular instant of time k_1 at which $x_1(k_1) = 0.4$ and $x_2(k_1) = -0.8$, the instantaneous fuzzy basis function values are

$$\xi_1(k_1) = \frac{\mu_1(k_1)}{\sum_{i=1}^{4} \mu_i(k_1)} = \frac{0.25}{0.25 + 0.0272 + 0.7578 + 0.0825} = 0.2237$$

$$\xi_2(k_1) = \frac{\mu_2(k_1)}{\sum_{i=1}^{4} \mu_i(k_1)} = \frac{0.0272}{0.25 + 0.0272 + 0.7578 + 0.0825} = 0.0243$$

$$\xi_3(k_1) = \frac{\mu_3(k_1)}{\sum_{i=1}^{4} \mu_i(k_1)} = \frac{0.7578}{0.25 + 0.0272 + 0.7578 + 0.0825} = 0.6782$$

$$\xi_4(k_1) = \frac{\mu_4(k_1)}{\sum_{i=1}^{4} \mu_i(k_1)} = \frac{0.0825}{0.25 + 0.0272 + 0.7578 + 0.0825} = 0.0738$$

Then the system at time k_1 is described by the difference equation

$$x(k+1) = A(k_1)x(k) + b(k_1)u(k)$$

where

$$\begin{aligned} A(k_1) &= \xi_1(k_1)A_1 + \xi_2(k_1)A_2 + \xi_3(k_1)A_3 + \xi_4(k_1)A_4 \\ &= 0.2237\begin{bmatrix} 0.1 & 0 \\ -0.2 & 0.2 \end{bmatrix} + 0.0243\begin{bmatrix} 0 & 0.1 \\ 1 & -0.2 \end{bmatrix} + 0.6782\begin{bmatrix} -0.3 & 1 \\ 0 & 0.2 \end{bmatrix} \\ &\quad + 0.0738\begin{bmatrix} 0.1 & 1 \\ 0.2 & 0.5 \end{bmatrix} = \begin{bmatrix} -0.1737 & 0.7544 \\ -0.0056 & 0.2124 \end{bmatrix} \end{aligned}$$

and

$$\begin{aligned} b(k_1) &= \xi_1(k_1)b_1 + \xi_2(k_1)b_2 + \xi_3(k_1)b_3 + \xi_4(k_1)b_4 \\ &= 0.2237\begin{bmatrix} 0 \\ 1 \end{bmatrix} + 0.0243\begin{bmatrix} 1 \\ 1 \end{bmatrix} + 0.6782\begin{bmatrix} -1 \\ 1 \end{bmatrix} + 0.0738\begin{bmatrix} 0 \\ -1 \end{bmatrix} = \begin{bmatrix} -0.06538 \\ 0.8524 \end{bmatrix} \end{aligned}$$

To summarize, the system at time k_1 is described by

$$\begin{bmatrix} x_1(k+1) \\ x_2(k+1) \end{bmatrix} = \begin{bmatrix} -0.1737 & 0.7544 \\ -0.0056 & 0.2124 \end{bmatrix}\begin{bmatrix} x_1(k) \\ x_2(k) \end{bmatrix} + \begin{bmatrix} -0.06538 \\ 0.8524 \end{bmatrix} u(k)$$

Note that this describes the system at only one time k_1. As k changes, $A(k)$ and $b(k)$ change.

This system was simulated with $x_1(0) = x_2(0) = 0$, a sampling time of $\Delta t = 0.2$ s, and $u(t) = 3\sin(\pi t)$, where $t = k\Delta t$ and k is an integer. The resulting state trajectories

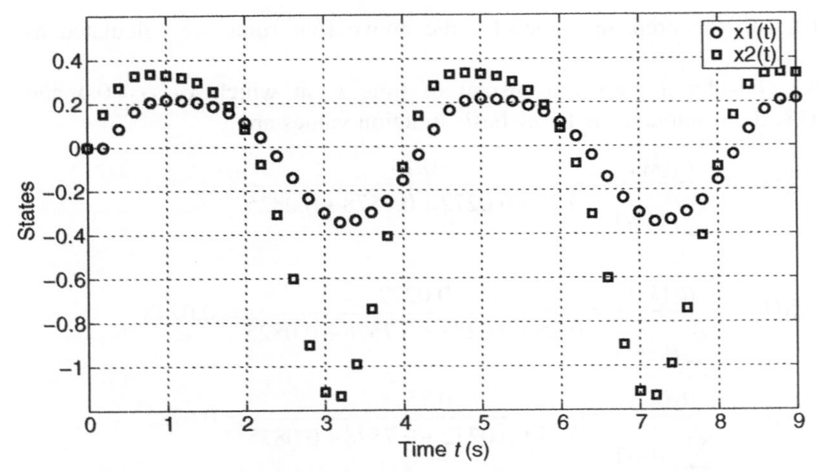

Figure 6.7. Input and corresponding states for discrete-time system created by a four-rule T–S fuzzy system.

are shown in Figure 6.7. The nonlinear nature of the system can be seen in this response. The Matlab simulation program producing Figure 6.7 is given in the Appendix.

6.4 TAKAGI–SUGENO FUZZY SYSTEMS AS INTERPOLATORS BETWEEN DISCRETE-TIME DYNAMIC SYSTEMS DESCRIBED BY INPUT–OUTPUT DIFFERENCE EQUATIONS

The above method can be used to create a nonlnear time-varying discrete-time dynamic system described by an input–output difference equation. This form of system model is particularly useful when system identification techniques are being used because an unknown system is often identified in the form of a discrete-time input–output difference equation rather than state-space equations as above. If state-space equations are desired (e.g., for parallel distributed pole placement), they can always be obtained from input–output difference equations via well-known canonical forms.

Consider a T–S fuzzy system with R rules of the form:

$$\begin{aligned} R_i:\quad & \text{If } y(k) \text{ is } P_1^K \text{ and } y(k-1) \text{ is } P_2^L \text{ and} \ldots \text{and } y(k-n+1) \text{ is } P_n^M\text{, then} \\ & y^i(k+1) = \alpha_{i,1}y(k)+\alpha_{i,2}y(k-1)+\cdots+\alpha_{i,n}y(k-n+1)+ \\ & \qquad \beta_{i,1}u(k)+\beta_{i,2}u(k-1)+\cdots+\beta_{i,m}u(k-m+1) \end{aligned} \tag{6.15}$$

where $y(k)$, … , $y(k-n+1)$, are present and past plant outputs and $u(k)$, … , $u(k-m+1)$ are present and past plant inputs. This way of describing a discrete-time system is commonly used in system identification techniques.

This fuzzy system produces a time-varying discrete-time system described by the difference equation:

$$y(k+1)=\frac{\sum_{i=1}^{R}\mu_i(y(k),\ldots,y(k-n+1))\,y^i(k+1)}{\sum_{i=1}^{R}\mu_i(y(k),\ldots,y(k-n+1))} \tag{6.16}$$

where $\mu_i(y(k), \ldots, y(k-n+1))$ is the premise membership value of Rule i. Therefore, the discrete-time system produced by the T–S fuzzy system is

$$\begin{aligned} y(k+1)=\alpha_1 y(k)+\alpha_2 y(k-1)+\cdots+\alpha_n y(k-n+1)+\\ \beta_1 u(k)+\beta_2 u(k-1)+\cdots+\beta_m u(k-m+1)\end{aligned} \tag{6.17}$$

where

$$\begin{aligned}
\alpha_1 &= \alpha_{1,1}\xi_1(k)+\alpha_{2,1}\xi_2(k)+\cdots+\alpha_{R,1}\xi_R(k)\\
\alpha_2 &= \alpha_{1,2}\xi_1(k)+\alpha_{2,2}\xi_2(k)+\cdots+\alpha_{R,2}\xi_R(k)\\
&\vdots\\
\alpha_n &= \alpha_{1,n}\xi_1(k)+\alpha_{2,n}\xi_2(k)+\cdots+\alpha_{R,n}\xi_R(k)\\
\beta_1 &= \beta_{1,1}\xi_1(k)+\beta_{2,1}\xi_2(k)+\cdots+\beta_{R,1}\xi_R(k)\\
\beta_2 &= \beta_{1,2}\xi_1(k)+\beta_{2,2}\xi_2(k)+\cdots+\beta_{R,2}\xi_R(k)\\
&\vdots\\
\beta_m &= \beta_{1,m}\xi_1(k)+\beta_{2,m}\xi_2(k)+\cdots+\beta_{R,m}\xi_R(k)
\end{aligned}$$

and the fuzzy basis functions are given by

$$\xi_j(k)=\frac{\mu_j(k)}{\sum_{i=1}^{R}\mu_i(k)},\ j=1,\ldots,R$$

EXAMPLE 6.4

Consider a two-input T–S fuzzy system with rules

1. If $y(k)$ is P_1^1 and $y(k-1)$ is P_2^1, then

$$y^1(k+1)=0.5y(k)-0.5y(k-1)+u(k)+0.6u(k-1).$$

2. If $y(k)$ is P_1^1 and $y(k-1)$ is P_2^2, then

$$y^2(k+1)=0.4y(k)-0.8y(k-1)+1.2u(k)-u(k-1).$$

3. If $y(k)$ is P_1^2 and $y(k-1)$ is P_2^1, then

$$y^3(k+1)=0.2y(k)+0.5y(k-1)+1.5u(k)-0.7u(k-1).$$

4. If $y(k)$ is P_1^2 and $y(k-1)$ is P_2^2, then

$$y^4(k+1)=0.8y(k)-0.6y(k-1)-1.5u(k)+u(k-1).$$

Let the fuzzy sets P_1^1, P_1^2, P_2^1, and P_2^2 be characterized by the Gaussian membership functions shown in Figures 6.1 and 6.2. Then for an arbitrary input (x_1, x_2) the premise values for the above four rules (μ_i, $i = 1, \ldots, 4$) are calculated as in (6.6).

If we consider a particular instant of time k_1 at which $y(k_1) = 0.4$ and $y(k_1 - 1) = -0.8$, the instantaneous fuzzy basis function values are the same as in Example 6.3. Then the system at time k_1 is described by the input-output difference equation

$$y(k+1) = \alpha_1(k_1)y(k) + \alpha_2(k_1)y(k-1) + \beta_1(k_1)u(k) + \beta_2(k_1)u(k-1)$$

where

$$\begin{aligned}
\alpha_1(k_1) &= 0.5\xi_1(k_1) + 0.4\xi_2(k_1) + 0.2\xi_3(k_1) + 0.8\xi_4(k_1) = 0.3163 \\
\alpha_2(k_1) &= -0.5\xi_1(k_1) - 0.8\xi_2(k_1) + 0.5\xi_3(k_1) - 0.6\xi_4(k_1) = 0.1635 \\
\beta_1(k_1) &= 1\xi_1(k_1) + 1.2\xi_2(k_1) + 1.5\xi_3(k_1) - 1.5\xi_4(k_1) = 1.1595 \\
\beta_2(k_1) &= 0.6\xi_1(k_1) - 1\xi_2(k_1) - 0.7\xi_3(k_1) + 1\xi_4(k_1) = -0.2911
\end{aligned}$$

To summarize, the system at time k_1 is described by

$$y(k+1) = 0.3163y(k) + 0.1635y(k-1) + 1.1595u(k) - 0.2911u(k-1)$$

Note that this describes the system at only one time k_1. As k changes, α_1, α_2, β_1, and β_2 also change. A simulation of this system with a sampling time of $\Delta t = 1$ s, $y(0) = y(-1) = 0$, and $u(t) = \sin(0.05\pi t)$, where $t = k\Delta t$ yields the trajectory shown in Figure 6.8. The nonlinear nature of the system can be seen in this response. The Matlab program producing Figure 6.8 is given in the Appendix.

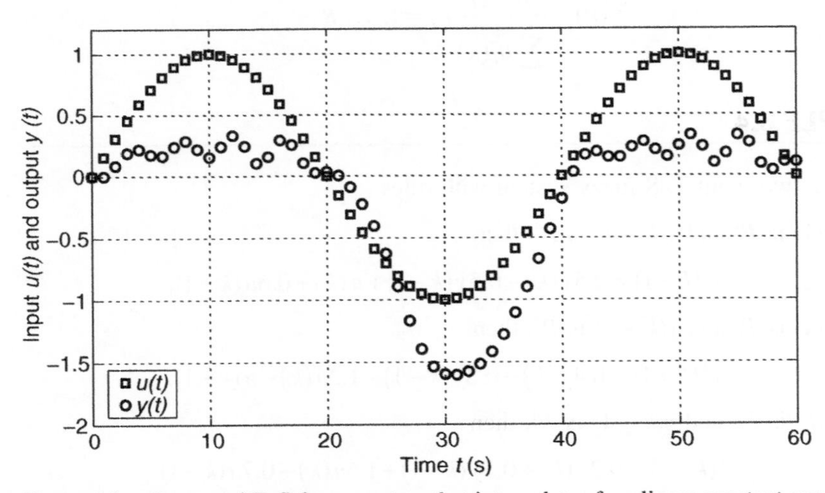

Figure 6.8. Output of T–S fuzzy system that interpolates four linear constant-coefficient discrete-time dynamic systems to produce a nonlinear discrete-time dynamic system.

6.5 SUMMARY

This chapter introduces **Takagi–Sugeno (T–S) fuzzy systems**, which are generalizations of Mamdani fuzzy systems (Chapter 3). The consequents of T–S systems are general mathematical functions rather than fuzzy sets. T–S fuzzy systems make identification of nonlinear systems (Chapter 9) possible, as well as adaptive fuzzy control (Chapter 10). They also enable stability proofs for closed-loop systems controlled by T–S fuzzy controllers (Chapter 7).

The three types of T–S fuzzy systems considered in this book are those with consequents that are (1) affine functions, (2) normal form state-space models, or (3) input–output difference equation models.

If the consequents are **affine functions**, the T–S fuzzy system realizes a static nonlinear function, as in Example 6.1 and Figure 6.3. It was seen that this type of fuzzy system interpolates the hyperplanes in the individual consequents inside the effective universe of discourse. This type of fuzzy system will be utilized in Chapters 9 and 10 to identify nonlinear plants in the form of (5.3) for use in adaptive fuzzy controllers.

If the consequents are **normal form state-space models**, the T–S fuzzy system realizes a nonlinear time-varying state-space system, as in Sections 6.2, 6.3, and Examples 6.2 and 6.3. This type of fuzzy system will be utilized in Chapter 7 to design parallel distributed state-feedback controllers for nonlinear systems.

If the consequents are **input–output difference equation models**, the T–S fuzzy system realizes a nonlinear/time-varying system described by an input–output difference equation. This type of fuzzy system is utilized in Chapter 7 for parallel distributed tracking and model reference control, and again in Chapter 10 for adaptive versions of these.

EXERCISES

6.1 In Example 6.1, let $x_1 = -0.5$, $x_2 = -0.5$ and determine y^{crisp}.

6.2 Consider the T–S fuzzy system with rule base

1. If x_1 is P_1^1 and x_2 is P_2^1, then $q^1 = 1 + x_1^2$.
2. If x_1 is P_1^1 and x_2 is P_2^2, then $q^2 = x_1 - x_2$.
3. If x_1 is P_1^1 and x_2 is P_2^3, then $q^3 = \cos x_1 - 2x_2$.
4. If x_1 is P_1^2 and x_2 is P_2^1, then $q^4 = -2 - x_1$.
5. If x_1 is P_1^2 and x_2 is P_2^2, then $q^5 = x_2$.
6. If x_1 is P_1^2 and x_2 is P_2^3, then $q^6 = 2x_1 + x_2$.
7. If x_1 is P_1^3 and x_2 is P_2^1, then $q^7 = -1 - 2x_1x_2$.
8. If x_1 is P_1^3 and x_2 is P_2^2, then $q^8 = -2 - x_1 + 0.5x_2$.

9. If x_1 is P_1^3 and x_2 is P_2^3, then $q^9 = 1 - \sin x_1$.

and input fuzzy sets characterized by the memberships in Figure 6.9.

(a) Find the crisp output if $x_1 = -0.75$, $x_2 = 0.2$ using *product* T-norm.

(b) Repeat part (a) using *minimum* T-norm.

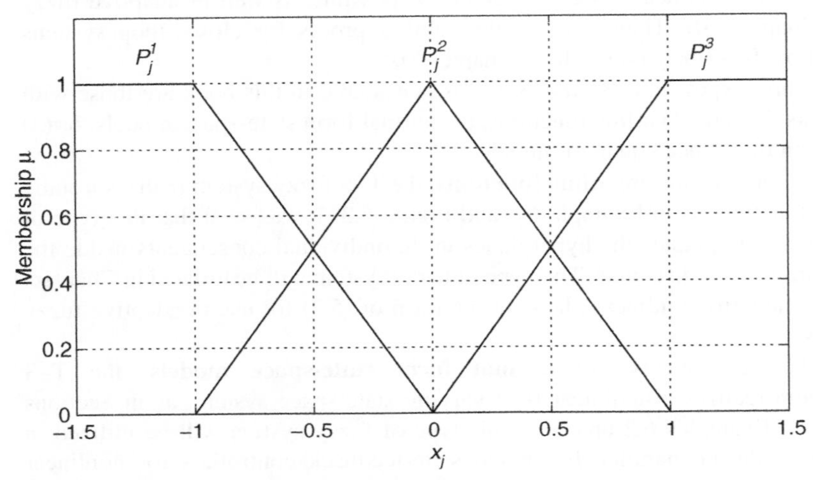

Figure 6.9. Input fuzzy sets for Problems 6.2 and 6.5, $j = 1, 2$.

6.3 Plot the characteristic input–output surface of the system of Problem 6.2 in the range $-1 \le x_1 \le 1$, $-1 \le x_2 \le 1$.

6.4 Consider the T–S fuzzy system with rule base

1. If x_1 is P_1^1 and x_2 is P_2^1, then $q^1 = 1 + x_1 + x_2$.

2. If x_1 is P_1^1 and x_2 is P_2^2, then $q^2 = x_1 - x_2$.

3. If x_1 is P_1^1 and x_2 is P_2^3, then $q^3 = x_1 - 2x_2$.

4. If x_1 is P_1^2 and x_2 is P_2^1, then $q^4 = -2 - x_1$.

5. If x_1 is P_1^2 and x_2 is P_2^2, then $q^5 = x_2$.

6. If x_1 is P_1^2 and x_2 is P_2^3, then $q^6 = 2x_1 + x_2$.

7. If x_1 is P_1^3 and x_2 is P_2^1, then $q^7 = -1 - 2x_2$.

8. If x_1 is P_1^3 and x_2 is P_2^2, then $q^8 = -2 - x_1 + 0.5x_2$.

9. If x_1 is P_1^3 and x_2 is P_2^3, then $q^9 = 1 - x_1$.

and input fuzzy sets characterized by the memberships in Figure 6.10.

(a) Find the crisp output if $x_1 = 3.5$, $x_2 = 0.5$ using *product* T-norm.

(b) Plot the characteristic input–output surface of the system in the range $-0.5 \le x_1 \le 4.5$, $-0.5 \le x_2 \le 4.5$.

6.5 In Example 6.2, let $x_1(t_1) = -0.5$, $x_2(t_1) = -0.5$ and determine the continuous-time state-space differential equation that describes the fuzzy system at time t_1.

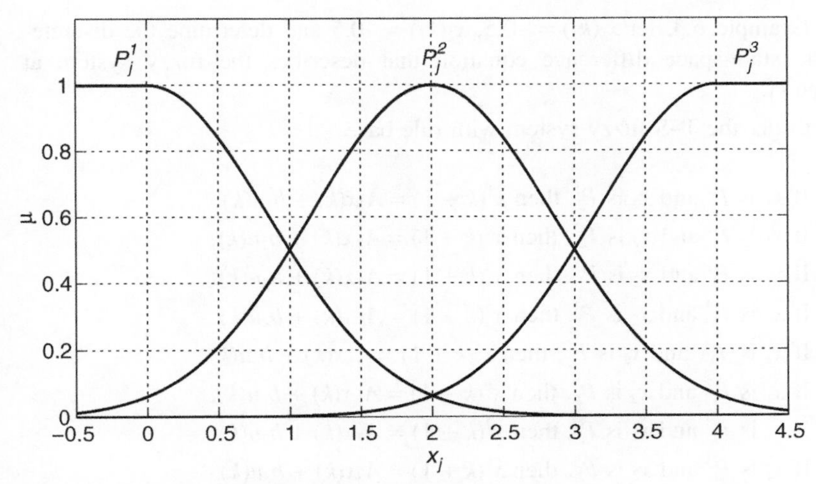

Figure 6.10. Input fuzzy sets for Problem 6.3, $j = 1, 2$.

6.6 Consider the T–S fuzzy system with rule base

1. If x_1 is P_1^1 and x_2 is P_2^1, then $\dot{x}^1(t) = A_1x(t) + b_1u(t)$.
2. If x_1 is P_1^1 and x_2 is P_2^2, then $\dot{x}^2(t) = A_2x(t) + b_2u(t)$.
3. If x_1 is P_1^1 and x_2 is P_2^3, then $\dot{x}^3(t) = A_3x(t) + b_3u(t)$.
4. If x_1 is P_1^2 and x_2 is P_2^1, then $\dot{x}^4(t) = A_4x(t) + b_4u(t)$.
5. If x_1 is P_1^2 and x_2 is P_2^2, then $\dot{x}^5(t) = A_5x(t) + b_5u(t)$.
6. If x_1 is P_1^2 and x_2 is P_2^3, then $\dot{x}^6(t) = A_6x(t) + b_6u(t)$.
7. If x_1 is P_1^3 and x_2 is P_2^1, then $\dot{x}^7(t) = A_7x(t) + b_7u(t)$.
8. If x_1 is P_1^3 and x_2 is P_2^2, then $\dot{x}^8(t) = A_8x(t) + b_8u(t)$.
9. If x_1 is P_1^3 and x_2 is P_2^3, then $\dot{x}^9(t) = A_9x(t) + b_9u(t)$.

with $A_1 = \begin{bmatrix} 0 & 1 \\ -2 & -2 \end{bmatrix}$, $b_1 = \begin{bmatrix} 0 \\ 2 \end{bmatrix}$, $A_2 = \begin{bmatrix} 0 & 1 \\ -1 & -2 \end{bmatrix}$, $b_2 = \begin{bmatrix} 2 \\ 2 \end{bmatrix}$

$A_3 = \begin{bmatrix} 0 & 1 \\ -3 & -2 \end{bmatrix}$, $b_3 = \begin{bmatrix} -2 \\ 2 \end{bmatrix}$, $A_4 = \begin{bmatrix} 1 & 1 \\ -2 & -2 \end{bmatrix}$, $b_4 = \begin{bmatrix} 0 \\ -2 \end{bmatrix}$

$A_5 = \begin{bmatrix} 0 & 1 \\ -2 & -1 \end{bmatrix}$, $b_5 = \begin{bmatrix} 0 \\ 1 \end{bmatrix}$, $A_6 = \begin{bmatrix} 0 & 1 \\ -1 & -2 \end{bmatrix}$, $b_6 = \begin{bmatrix} 1 \\ 1 \end{bmatrix}$

$A_7 = \begin{bmatrix} 0 & 1 \\ -2 & -2 \end{bmatrix}$, $b_7 = \begin{bmatrix} -1 \\ 1 \end{bmatrix}$, $A_8 = \begin{bmatrix} -1 & -1 \\ 1 & 0 \end{bmatrix}$, $b_8 = \begin{bmatrix} 0 \\ -1 \end{bmatrix}$, $A_9 = \begin{bmatrix} 0 & 1 \\ 2 & -2 \end{bmatrix}$, $b_9 = \begin{bmatrix} 0 \\ 1 \end{bmatrix}$

and $u(t) = 3\sin(\pi t)$. Let the input fuzzy sets be characterized by the memberships in Figure 6.9. Simulate the system and plot $x_1(t)$, $x_2(t)$, $0 \le t < 8$ as in Figure 6.6.

6.7 In Example 6.3, let $x_1(k_1) = -0.5$, $x_2(k_1) = -0.5$ and determine the discrete-time state-space difference equation that describes the fuzzy system at time k_1.

6.8 Consider the T–S fuzzy system with rule base

1. If x_1 is P_1^1 and x_2 is P_2^1, then $x^1(k+1) = A_1x(k) + b_1u(k)$.
2. If x_1 is P_1^1 and x_2 is P_2^2, then $x^2(k+1) = A_2x(k) + b_2u(k)$.
3. If x_1 is P_1^1 and x_2 is P_2^3, then $x^3(k+1) = A_3x(k) + b_3u(k)$.
4. If x_1 is P_1^2 and x_2 is P_2^1, then $x^4(k+1) = A_4x(k) + b_4u(k)$.
5. If x_1 is P_1^2 and x_2 is P_2^2, then $x^5(k+1) = A_5x(k) + b_5u(k)$.
6. If x_1 is P_1^2 and x_2 is P_2^3, then $x^6(k+1) = A_6x(k) + b_6u(k)$.
7. If x_1 is P_1^3 and x_2 is P_2^1, then $x^7(k+1) = A_7x(k) + b_7u(k)$.
8. If x_1 is P_1^3 and x_2 is P_2^2, then $x^8(k+1) = A_8x(k) + b_8u(k)$.
9. If x_1 is P_1^3 and x_2 is P_2^3, then $x^9(k+1) = A_9x(k) + b_9u(k)$.

with

$$A_1 = \begin{bmatrix} 0.1 & 0 \\ -0.2 & 0.2 \end{bmatrix}, b_1 = \begin{bmatrix} 0 \\ 1 \end{bmatrix}, A_2 = \begin{bmatrix} 0 & 0.1 \\ 1 & -0.2 \end{bmatrix}, b_2 = \begin{bmatrix} 0.1 \\ 1 \end{bmatrix}$$

$$A_3 = \begin{bmatrix} -0.3 & 1 \\ 0 & 0.2 \end{bmatrix}, b_3 = \begin{bmatrix} -0.1 \\ 1 \end{bmatrix}, A_4 = \begin{bmatrix} 0.1 & 1 \\ 0.2 & 0.5 \end{bmatrix}, b_4 = \begin{bmatrix} 0 \\ -1 \end{bmatrix}$$

$$A_5 = \begin{bmatrix} 0.1 & 0 \\ -0.1 & -0.2 \end{bmatrix}, b_5 = \begin{bmatrix} 0 \\ 1 \end{bmatrix}, A_6 = \begin{bmatrix} 0 & -0.1 \\ 1 & 0.2 \end{bmatrix}, b_6 = \begin{bmatrix} 0.1 \\ 1 \end{bmatrix}$$

$$A_7 = \begin{bmatrix} 0.3 & 1 \\ 0 & 0.2 \end{bmatrix}, b_7 = \begin{bmatrix} 0.1 \\ 1 \end{bmatrix}, A_8 = \begin{bmatrix} 0.1 & 1 \\ -0.2 & 0 \end{bmatrix}, b_8 = \begin{bmatrix} 0 \\ 1 \end{bmatrix}, A_9 = \begin{bmatrix} -0.1 & 0 \\ 0.2 & 0.2 \end{bmatrix}, b_9 = \begin{bmatrix} 1 \\ 0.1 \end{bmatrix}$$

and $u(t) = 3\sin(\pi t)$ where $t = k\Delta t$ and k is an integer. Let $\Delta t = 0.5$ s. Let the input fuzzy sets be characterized by the memberships in Figure 6.9. Simulate the system and plot $x_1(t)$, $x_2(t)$, $0 \le t < 9$ as in Figure 6.7.

6.9 In Example 6.4, let $y(k_1) = -0.5$, $y(k_1 - 1) = -0.5$ and determine the discrete-time input–output difference equation that describes the fuzzy system at time k_1.

6.10 Consider the T–S fuzzy system with rule base

1. If $y(k)$ is P_1^1 and $y(k-1)$ is P_2^1, then $y^1(k+1) = \alpha^1(q^{-1})y(k) + \beta^1(q^{-1})u(k)$.
2. If $y(k)$ is P_1^1 and $y(k-1)$ is P_2^2, then $y^2(k+1) = \alpha^2(q^{-1})y(k) + \beta^2(q^{-1})u(k)$.
3. If $y(k)$ is P_1^1 and $y(k-1)$ is P_2^3, then $y^3(k+1) = \alpha^3(q^{-1})y(k) + \beta^3(q^{-1})u(k)$.
4. If $y(k)$ is P_1^2 and $y(k-1)$ is P_2^1, then $y^4(k+1) = \alpha^4(q^{-1})y(k) + \beta^4(q^{-1})u(k)$.
5. If $y(k)$ is P_1^2 and $y(k-1)$ is P_2^2, then $y^5(k+1) = \alpha^5(q^{-1})y(k) + \beta^5(q^{-1})u(k)$.
6. If $y(k)$ is P_1^2 and $y(k-1)$ is P_2^3, then $y^6(k+1) = \alpha^6(q^{-1})y(k) + \beta^6(q^{-1})u(k)$.

7. If $y(k)$ is P_1^3 and $y(k-1)$ is P_2^1, then $y^7(k+1) = \alpha^7(q^{-1})y(k) + \beta^7(q^{-1})u(k)$.
8. If $y(k)$ is P_1^3 and $y(k-1)$ is P_2^2, then $y^8(k+1) = \alpha^8(q^{-1})y(k) + \beta^8(q^{-1})u(k)$.
9. If $y(k)$ is P_1^3 and $y(k-1)$ is P_2^3, then $y^9(k+1) = \alpha^9(q^{-1})y(k) + \beta^9(q^{-1})u(k)$.

where

$$\alpha^1(q^{-1}) = 0.5 - 0.5q^{-1} \qquad \beta^1(q^{-1}) = 1 + 0.6q^{-1}$$
$$\alpha^2(q^{-1}) = 0.4 - 0.8q^{-1} \qquad \beta^2(q^{-1}) = 1.2 - q^{-1}$$
$$\alpha^3(q^{-1}) = 0.2 + 0.5q^{-1} \qquad \beta^3(q^{-1}) = 1.5 - 0.7q^{-1}$$
$$\alpha^4(q^{-1}) = 0.8 - 0.6q^{-1} \qquad \beta^4(q^{-1}) = -1.5 + q^{-1}$$
$$\alpha^5(q^{-1}) = -0.5 - 0.2q^{-1} \qquad \beta^5(q^{-1}) = 1 + 0.2q^{-1}$$
$$\alpha^6(q^{-1}) = 0.8 + 0.5q^{-1} \qquad \beta^6(q^{-1}) = 0.5 + 0.3q^{-1}$$
$$\alpha^7(q^{-1}) = 0.9 - 0.3q^{-1} \qquad \beta^7(q^{-1}) = 1 + 0.7q^{-1}$$
$$\alpha^8(q^{-1}) = 0.7 - 0.5q^{-1} \qquad \beta^8(q^{-1}) = 1 + 0.8q^{-1}$$
$$\alpha^9(q^{-1}) = 0.5 + 0.4q^{-1} \qquad \beta^9(q^{-1}) = 1 + 0.3q^{-1}$$

Let $u(t) = \sin(0.05\pi t)$, where $t = k\Delta t$ and k is an integer. Let $\Delta t = 1$ s and $y(0) = y(-1) = 0$. Let the input fuzzy sets be characterized by the memberships in Figure 6.11 (note that adjacent memberships cross at 0.3). Simulate the system and plot $y(t)$, $0 \leq t < 60$ as in Figure 6.8.

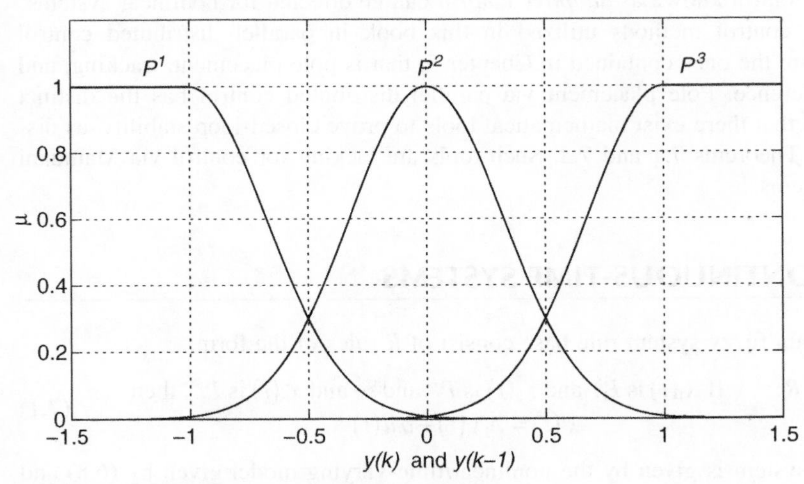

Figure 6.11. Input fuzzy sets for Problem 6.10.

CHAPTER 7

PARALLEL DISTRIBUTED CONTROL WITH TAKAGI–SUGENO FUZZY SYSTEMS

If a plant has a model in the form of a Takagi–Sugeno (T–S) fuzzy system that interpolates linear dynamic systems, which we call the *plant fuzzy system*, one approach to stabilizing it is called *parallel distributed control* (PDC) [9,17,29]. The basic philosophy of PDC is to create a *controller fuzzy system* with rule premises identical to those of the plant fuzzy system. In the controller fuzzy system, each rule's consequent is a control law designed to control the linear system in the corresponding consequent of the plant fuzzy system. The overall control law is thus a weighted average of the individual control laws, just as the overall nonlinear system is a weighted average of the linear systems in each consequent of the plant fuzzy system.

Parallel distributed control is important because it constitutes a method of controlling nonlinear systems. Furthermore, as will be seen in Chapter 10, a nonlinear system can sometimes be identified online as a T–S fuzzy system. If a parallel distributed controller can then be designed based on this identification, a type of real-time control known as *adaptive control* can be effected for nonlinear systems.

The control methods utilized in this book in parallel distributed control schemes are the ones contained in Chapter 5, that is pole placement, tracking, and model reference. Pole placement via parallel distributed control has the distinct advantage that there exist mathematical tools to prove closed-loop stability, as discussed in Theorems 7.1 and 7.2. Such tools are lacking for control via Mamdani fuzzy systems.

7.1 CONTINUOUS-TIME SYSTEMS

Let the plant fuzzy system rule base consist of R rules of the form:

$$R_i: \quad \text{If } x_1(t) \text{ is } P_1^K \text{ and } x_2(t) \text{ is } P_2^L \text{ and} \cdots \text{and } x_n(t) \text{ is } P_n^M, \text{ then} \\ \dot{x}^i(t) = A_i x(t) + b_i u(t) \tag{7.1}$$

Then the system is given by the nonlinear/time-varying model given by (6.8) and (6.9).

Fuzzy Control and Identification, By John H. Lilly

The parallel distributed pole placement controller for this system is another fuzzy system with R rules of the form:

$$R_i \qquad \text{If } x_1(t) \text{ is } P_1^K \text{ and } x_2(t) \text{ is } P_2^L \text{ and} \cdots \text{and } x_n(t) \text{ is } P_n^M \text{, then} \quad u^i(t) = -k_i x(t) \tag{7.2}$$

where the eigenvalues of $A_i - b_i k_i$ are as desired in the left half-plane (see Section 5.3.1). The resulting control law is

$$u(t) = -\left[\sum_{i=1}^{R} k_i \xi_i(t)\right] x(t) \tag{7.3}$$

where $\xi_i(t)$ are the fuzzy basis functions defined in (6.4).

Not every PDC controller produces a stable closed-loop system, despite the fact that each consequent system in the plant fuzzy system is stabilized by the control law in the corresponding consequent of the controller fuzzy system. Nevertheless, we have the following stability result for pole placement via parallel distributed control.

THEOREM 7.1 (CONTINUOUS-TIME PARALLEL DISTRIBUTED POLE PLACEMENT)

In the parallel distributed pole placement scheme described by (7.1)–(7.3), assume each consequent system of the plant fuzzy system is controllable. Then, $\underline{x} = \underline{0}$ is globally asymptotically stable if there exists a positive definite symmetric matrix G such that for all $i = 1, \ldots, R$ and $j = 1, \ldots, R$,

$$G(A_i - b_i k_j) + (A_i - b_i k_j)^{\mathrm{T}} G < 0 \tag{7.4}$$

where $M < 0$ means that the matrix M is negative definite.

The proof follows directly from Lyapunov stability theory. Note that the inequality (7.4) must be satisfied by *one* matrix G for all combinations of i and j. The inequality (7.4) is called a *continuous-time linear matrix inequality* (LMI) [29–31]. LMIs, such as (7.4), are difficult to solve by hand. Therefore, there are various computational tools to solve them.

EXAMPLE 7.1

Consider a nonlinear two-state dynamic system described by the T–S fuzzy system

1. If x_1 is P_1^1 and x_2 is P_2^1, then $\dot{x}^1(t) = A_1 x(t) + b_1 u(t)$.
2. If x_1 is P_1^1 and x_2 is P_2^2, then $\dot{x}^2(t) = A_2 x(t) + b_2 u(t)$.
3. If x_1 is P_1^2 and x_2 is P_2^1, then $\dot{x}^3(t) = A_3 x(t) + b_3 u(t)$.
4. If x_1 is P_1^2 and x_2 is P_2^2, then $\dot{x}^4(t) = A_4 x(t) + b_4 u(t)$.

where

$$A_1 = \begin{bmatrix} 0 & 1 \\ -2 & 2 \end{bmatrix}, b_1 = \begin{bmatrix} 0 \\ 1 \end{bmatrix}, A_2 = \begin{bmatrix} 0 & 1 \\ -3 & 2 \end{bmatrix}, b_2 = \begin{bmatrix} 0 \\ 1 \end{bmatrix}$$
$$A_3 = \begin{bmatrix} 0 & 1 \\ -2 & 1 \end{bmatrix}, b_3 = \begin{bmatrix} 0 \\ 1 \end{bmatrix}, A_4 = \begin{bmatrix} 0 & 1 \\ -3 & 1 \end{bmatrix}, b_4 = \begin{bmatrix} 0 \\ 1 \end{bmatrix}$$

Notice that each consequent system is controllable and unstable. The eigenvalues of A_1 are $1 \pm j$, those of A_2 are $1 \pm j\sqrt{2}$, those of A_3 are $0.5 \pm j1.32$, and those of A_4 are $0.5 \pm j1.66$. Therefore, this T–S fuzzy system is nonlinear and unstable. To stabilize it, we design a parallel distributed pole placement controller with state feedback control laws to place the closed-loop eigenvalues of each system at $-1 \pm j$. The parallel distributed controller is given by

1. If $x_1(t)$ is P_1^1 and $x_2(t)$ is P_2^1, then $u^1(t) = -k_1 x(t)$.
2. If $x_1(t)$ is P_1^1 and $x_2(t)$ is P_2^2, then $u^2(t) = -k_2 x(t)$.
3. If $x_1(t)$ is P_1^2 and $x_2(t)$ is P_2^1, then $u^3(t) = -k_3 x(t)$.
4. If $x_1(t)$ is P_1^2 and $x_2(t)$ is P_2^2, then $u^4(t) = -k_4 x(t)$.

where $k_1 = [0 \quad 4]$, $k_2 = [-1 \quad 4]$, $k_3 = [0 \quad 3]$, and $k_4 = [-1 \quad 3]$. Then the control law applied to the plant is

$$u(t) = -[k_1\xi_1(t) + k_2\xi_2(t) + k_3\xi_3(t) + k_4\xi_4(t)]x(t) \tag{7.5}$$

The requirement (7.4) for stability amounts to 16 inequalities for this particular problem. These are

$$G(A_1 - b_1k_1) + (A_1 - b_1k_1)^T G < 0$$
$$G(A_1 - b_1k_2) + (A_1 - b_1k_2)^T G < 0$$
$$G(A_1 - b_1k_3) + (A_1 - b_1k_3)^T G < 0$$
$$G(A_1 - b_1k_4) + (A_1 - b_1k_4)^T G < 0$$
$$G(A_2 - b_2k_1) + (A_2 - b_2k_1)^T G < 0$$
$$G(A_2 - b_2k_2) + (A_2 - b_2k_2)^T G < 0$$
$$G(A_2 - b_2k_3) + (A_2 - b_2k_3)^T G < 0$$
$$G(A_2 - b_2k_4) + (A_2 - b_2k_4)^T G < 0$$
$$G(A_3 - b_3k_1) + (A_3 - b_3k_1)^T G < 0$$
$$G(A_3 - b_3k_2) + (A_3 - b_3k_2)^T G < 0$$
$$G(A_3 - b_3k_3) + (A_3 - b_3k_3)^T G < 0$$
$$G(A_3 - b_3k_4) + (A_3 - b_3k_4)^T G < 0$$
$$G(A_4 - b_4k_1) + (A_4 - b_4k_1)^T G < 0$$
$$G(A_4 - b_4k_2) + (A_4 - b_4k_2)^T G < 0$$
$$G(A_4 - b_4k_3) + (A_4 - b_4k_3)^T G < 0$$
$$G(A_4 - b_4k_4) + (A_4 - b_4k_4)^T G < 0$$

As stated above, all 16 of these inequalities must be satisfied by one positive definite symmetric matrix G.

It may be verified that the matrix

$$G = \begin{bmatrix} 1509 & 347 \\ 347 & 653 \end{bmatrix}$$

satisfies these 16 LMIs. This G was found with the aid of the Matlab *LMI Control Toolbox*. Therefore, we are assured that the above parallel distributed controller renders the point $\underline{x} = \underline{0}$ globally asymptotically stable in the above closed-loop system.

Let the fuzzy sets P_1^1, P_1^2, P_2^1, and P_2^2 be characterized by the membership functions shown in Figures 6.1 and 6.2. With initial conditions $x_1(0) = 1$, $x_2(0) = -1$, and with no feedback, the state trajectories of Figure 7.1 result. The trajectories increase unboundedly because the open-loop system is unstable. With the same initial conditions and parallel distributed pole placement feedback control law (7.5), the trajectories of Figure 7.2 result, indicating a stable closed-loop system. The Matlab program producing Figures 7.1 and 7.2 is given in the Appendix.

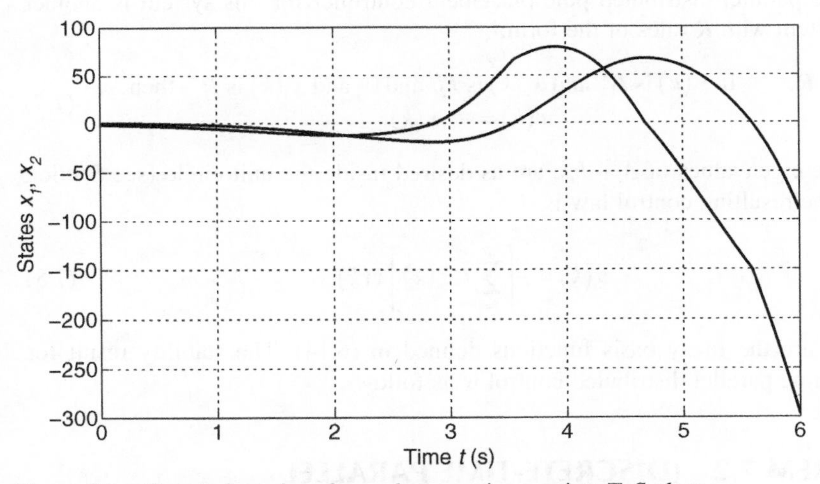

Figure 7.1. State trajectories of open-loop continuous-time T–S plant.

7.2 DISCRETE-TIME SYSTEMS

Let the plant fuzzy system rule base consist of R rules of the form:

$$R_i: \quad \text{If } x_1(k) \text{ is } P_1^K \text{ and } x_2(k) \text{ is } P_2^L \text{ and} \cdots \text{and } x_n(k) \text{ is } P_n^M\text{, then}$$
$$x^i(k+1) = A_i x(k) + b_i u(k) \tag{7.6}$$

Then the system obeys the nonlinear/time-varying model given by (6.12) and (6.13).

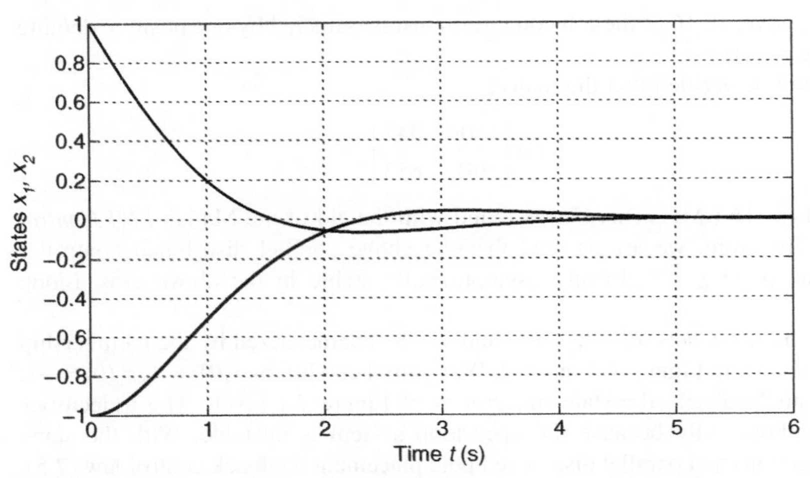

Figure 7.2. State trajectories of continuous-time T–S plant with parallel distributed pole placement.

The parallel distributed pole-placement controller for this system is another fuzzy system with R rules of the form:

$$R_i: \quad \text{If } x_1(k) \text{ is } P_1^K \text{ and } x_2(k) \text{ is } P_2^L \text{ and} \cdots \text{and } x_n(k) \text{ is } P_n^M, \text{ then} \quad u^i(k) = -k_i x(k) \tag{7.7}$$

where the eigenvalues of $A_i - b_i k_i$ are as desired inside the unit circle (see Section 5.3.1). The resulting control law is

$$u(k) = -\left[\sum_{i=1}^{R} k_i \xi_i(k)\right] x(k) \tag{7.8}$$

where ξ_i are the fuzzy basis functions defined in (6.14). The stability result for discrete-time parallel distributed control is as follows.

THEOREM 7.2 (DISCRETE-TIME PARALLEL DISTRIBUTED POLE PLACEMENT)

In the parallel distributed pole placement scheme described by (7.6)–(7.8), assume each consequent system of the plant fuzzy system is controllable. Then $\underline{x} = \underline{0}$ is globally asymptotically stable if there exists a positive definite symmetric matrix G such that for all $i = 1, \ldots, R$ and $j = 1, \ldots, R$,

$$(A_i - b_i k_j)^{\mathrm{T}} G (A_i - b_i k_j) - G < 0 \tag{7.9}$$

The inequality (7.9) is called a *discrete-time LMI*.

EXAMPLE 7.2

Consider a nonlinear dynamic system described by the T–S fuzzy system

1. If $x_1(k)$ is P_1^1 and $x_2(k)$ is P_2^1, then $x^1(k+1) = A_1x(k) + b_1u(k)$.
2. If $x_1(k)$ is P_1^1 and $x_2(k)$ is P_2^2, then $x^2(k+1) = A_2x(k) + b_2u(k)$.
3. If $x_1(k)$ is P_1^2 and $x_2(k)$ is P_2^1, then $x^3(k+1) = A_3x(k) + b_3u(k)$.
4. If $x_1(k)$ is P_1^2 and $x_2(k)$ is P_2^2, then $x^4(k+1) = A_4x(k) + b_4u(k)$.

where

$$A_1 = \begin{bmatrix} 0 & 1 \\ -1 & 1 \end{bmatrix}, b_1 = \begin{bmatrix} 0 \\ 1 \end{bmatrix}, A_2 = \begin{bmatrix} 1 & 1 \\ -1 & 2 \end{bmatrix}, b_2 = \begin{bmatrix} 0 \\ 1 \end{bmatrix}$$

$$A_3 = \begin{bmatrix} 0 & 1 \\ -0.2 & 1 \end{bmatrix}, b_3 = \begin{bmatrix} 0 \\ 1 \end{bmatrix}, A_4 = \begin{bmatrix} 1 & 1 \\ -0.3 & 1 \end{bmatrix}, b_4 = \begin{bmatrix} 0 \\ 1 \end{bmatrix}$$

The eigenvalues of A_1 are $0.5 \pm j0.866$, those of A_2 are $1.5 \pm j0.866$, those of A_3 are 0.2764, 0.7236, and those of A_4 are $1 \pm j0.5477$. Therefore, this T–S fuzzy system is nonlinear and unstable, because the eigenvalues of A_1, A_2, and A_4 have magnitudes that are not less than unity. To stabilize this system, we design a parallel distributed controller with state feedback control laws to place the closed-loop eigenvalues of each system at –0.5, 0.5. The parallel distributed controller is given by

1. If $x_1(k)$ is P_1^1 and $x_2(k)$ is P_2^1, then $u^1(k) = -k_1x(k)$.
2. If $x_1(k)$ is P_1^1 and $x_2(k)$ is P_2^2, then $u^2(k) = -k_2x(k)$.
3. If $x_1(k)$ is P_1^2 and $x_2(k)$ is P_2^1, then $u^3(k) = -k_3x(k)$.
4. If $x_1(k)$ is P_1^2 and $x_2(k)$ is P_2^2, then $u^4(k) = -k_4x(k)$.

where

$$k_1 = [-1.25 \quad 1], k_2 = [-0.25 \quad 3], k_3 = [-0.45 \quad 1], k_4 = [0.45 \quad 2]$$

Then the control law applied to the system is

$$u(k) = -[k_1\xi_1(k) + k_2\xi_2(k) + k_3\xi_3(k) + k_4\xi_4(k)]x(k) \tag{7.10}$$

Let the fuzzy sets P_1^1, P_1^2, P_2^1, and P_2^2 be characterized by the membership functions shown in Figures 6.1 and 6.2. With $\Delta t = 0.1$ s, initial conditions $x_1(0) = 1$, $x_2(0) = -1$, and with no feedback, the state trajectories of Figure 7.3 result. The trajectories increase unboundedly because the open-loop system is unstable. With the same initial conditions and parallel distributed pole placement feedback control law (7.10), the trajectories of Figure 7.4 result, indicating a stable closed-loop system. The Matlab program producing Figures 7.3 and 7.4 is given in the Appendix.

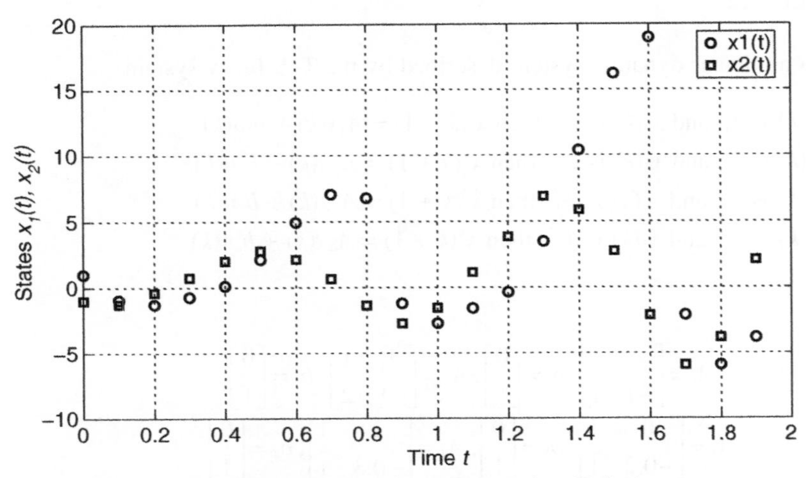

Figure 7.3. State trajectories of open-loop discrete-time T–S plant.

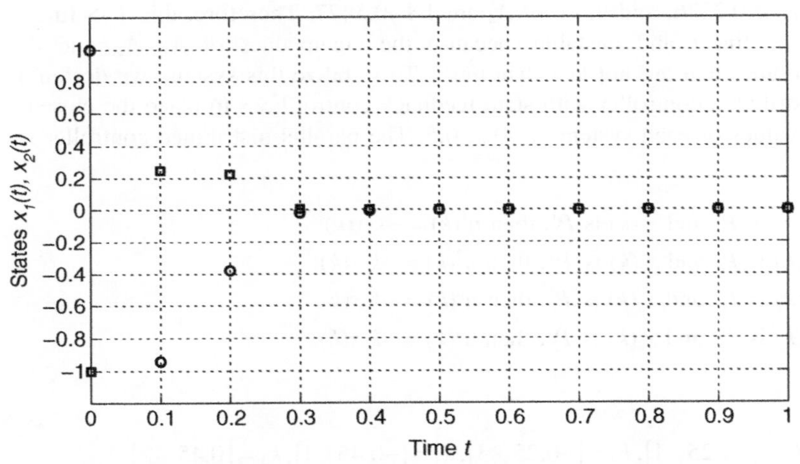

Figure 7.4. State trajectories of discrete-time T–S plant with parallel distributed pole placement.

7.3 PARALLEL DISTRIBUTED TRACKING CONTROL

The tracking control problem consists of determining a plant input such that the plant output tracks a desired reference signal. A parallel distributed-type controller can be designed to force the output of a nonlinear system modeled as a T–S fuzzy system to track a desired reference signal. The controller design is explained in Section 5.3.2 for a single linear plant. The parallel distributed tracking strategy is to apply this design procedure to every consequent plant in the plant fuzzy system. For tracking, the consequents in the plant fuzzy system are assumed to be in the form of (5.11), that is, input–output difference equations.

Let the plant fuzzy system rule base consist of R rules of the form:

$$R_i: \quad \text{If } y(k) \text{ is } P_1^K \text{ and } y(k-1) \text{ is } P_2^L \text{ and} \cdots \text{ and } y(k-n+1) \text{ is } P_n^M, \text{ then} \\ y^i(k+1) = \alpha_i(q^{-1})y(k) + \beta_i(q^{-1})u(k) \tag{7.11}$$

where

$$\alpha_i(q^{-1}) = a_{i,1} + a_{i,2}q^{-1} + \cdots + a_{i,n}q^{-(n-1)} \tag{7.12a}$$

$$\beta_i(q^{-1}) = b_{i,0} + b_{i,1}q^{-1} + \cdots + b_{i,m}q^{-m} = b_{i,0} + \beta_i'(q^{-1}) \tag{7.12b}$$

and q^{-1} is the backward shift operator defined by $q^{-1}y(k) = y(k-1)$.

The parallel distributed one-step-ahead tracking controller for this system is another fuzzy system with R rules of the form:

$$R_i: \quad \text{If } y(k) \text{ is } P_1^K \text{ and } y(k-1) \text{ is } P_2^L \text{ and} \cdots \text{ and } y(k-n+1) \text{ is } P_n^M, \text{ then} \\ u^i(k) = \frac{1}{b_{i,0}}\left[-\beta_i'(q^{-1})u(k) + r(k+1) - \alpha_i(q^{-1})y(k)\right] \tag{7.13}$$

The resulting control law is

$$u(k) = \sum_{i=1}^{R} u^i(k)\xi_i(k) \tag{7.14}$$

where $\xi_i(k)$, $i = 1, \ldots, R$ are the fuzzy basis functions defined in (6.14). This design procedure applies to n-step-ahead tracking problems with obvious modifications.

EXAMPLE 7.3

Consider the T–S fuzzy system with rule base:

1. If $y(k)$ is P_1^1 and $y(k-1)$ is P_2^1, then

$$y^1(k+1) = 1.5y(k) - 0.4y(k-1) + u(k) + 0.6u(k-1)$$

2. If $y(k)$ is P_1^1 and $y(k-1)$ is P_2^2, then

$$y^2(k+1) = 0.4y(k) - 1.8y(k-1) + 1.2u(k) + u(k-1)$$

3. If $y(k)$ is P_1^2 and $y(k-1)$ is P_2^1, then

$$y^3(k+1) = 1.2y(k) + 0.5y(k-1) + 1.5u(k) + 0.7u(k-1)$$

4. If $y(k)$ is P_1^2 and $y(k-1)$ is P_2^2, then

$$y^4(k+1) = 0.8y(k) - 1.6y(k-1) + 1.5u(k) + u(k-1)$$

The consequent systems are in the form of the consequent systems of (7.11) with

$$\alpha_1(q^{-1}) = 1.5 - 0.4q^{-1}, \alpha_2(q^{-1}) = 0.4 - 1.8q^{-1}, \alpha_3(q^{-1}) = 1.2 + 0.5q^{-1}$$

$$\alpha_4(q^{-1}) = 0.8 - 1.6q^{-1}, \beta_1(q^{-1}) = 1 + 0.6q^{-1}, \beta_2(q^{-1}) = 1.2 + q^{-1}$$

$$\beta_3(q^{-1}) = 1.5 + 0.7q^{-1},\ \beta_4(q^{-1}) = 1.5 + q^{-1},\ b_{1,0} = 1,\ \beta_1'(q^{-1}) = 0.6q^{-1},\ b_{2,0} = 1.2$$
$$\beta_2'(q^{-1}) = q^{-1},\ b_{3,0} = 1.5,\ \beta_3'(q^{-1}) = 0.7q^{-1},\ b_{4,0} = 1.5, \text{ and } \beta_4'(q^{-1}) = q^{-1}$$

Note that each consequent system is unstable because the polynomials

$$q^2[1 - q^{-1}\alpha_1(q^{-1})] = q^2 - 1.5q + 0.4,\ q^2[1 - q^{-1}\alpha_2(q^{-1})] = q^2 - 0.4q + 1.8$$
$$q^2[1 - q^{-1}\alpha_3(q^{-1})] = q^2 - 1.2q - 0.5, \text{ and } q^2[1 - q^{-1}\alpha_4(q^{-1})] = q^2 - 0.8q + 1.6$$

each have roots outside the unit circle.

As an example, consider the third consequent plant above, that is,

$$y^3(k+1) = 1.2y(k) + 0.5y(k-1) + 1.5u(k) + 0.7u(k-1)$$

For this plant, we have $\alpha_3(q^{-1}) = 1.2 + 0.5q^{-1}$, $\beta_3(q^{-1}) = 1.5 + 0.7q^{-1}$, $b_{3,0} = 1.5$, and $\beta_3'(q^{-1}) = 0.7q^{-1}$. Therefore, (5.31) yields

$$(1.5 + 0.7q^{-1})u^3(k) = r(k+1) - (1.2 + 0.5q^{-1})y(k)$$

to give the tracking control law

$$u^3(k) = \frac{1}{1.5}[-0.7u(k-1) - 1.2y(k) - 0.5y(k-1) + r(k+1)]$$

Using this design technique on all four consequent plants yields the parallel distributed one-step-ahead tracking controller:

1. If $y(k)$ is P_1^1 and $y(k-1)$ is P_2^1, then
$$u^1(k) = -0.6u(k-1) - 1.5y(k) + 0.4y(k-1) + r(k+1)$$
2. If $y(k)$ is P_1^1 and $y(k-1)$ is P_2^2, then
$$u^2(k) = \frac{1}{1.2}[-u(k-1) - 0.4y(k) + 1.8y(k-1) + r(k+1)]$$
3. If $y(k)$ is P_1^2 and $y(k-1)$ is P_2^1, then
$$u^3(k) = \frac{1}{1.5}[-0.7u(k-1) - 1.2y(k) - 0.5y(k-1) + r(k+1)]$$
4. If $y(k)$ is P_1^2 and $y(k-1)$ is P_2^2, then
$$u^4(k) = \frac{1}{1.5}[-u(k-1) - 0.8y(k) + 1.6y(k-1) + r(k+1)]$$

Let the input fuzzy sets P_1^1, P_1^2, P_2^1, and P_2^2 be defined as in Figure 6.4. Let the desired reference signal be $r(k) = 0.5\sin(0.2\pi k)$. With $\Delta t = 1$ second (hence $t = k\Delta t = k$), initial conditions $y(0) = 1$, $y(-1) = -1$, a plot of $y(t)$ together with $r(t)$ is shown in Figure 7.5 and the corresponding tracking error $y(t) - r(t)$ is shown in Figure 7.6.

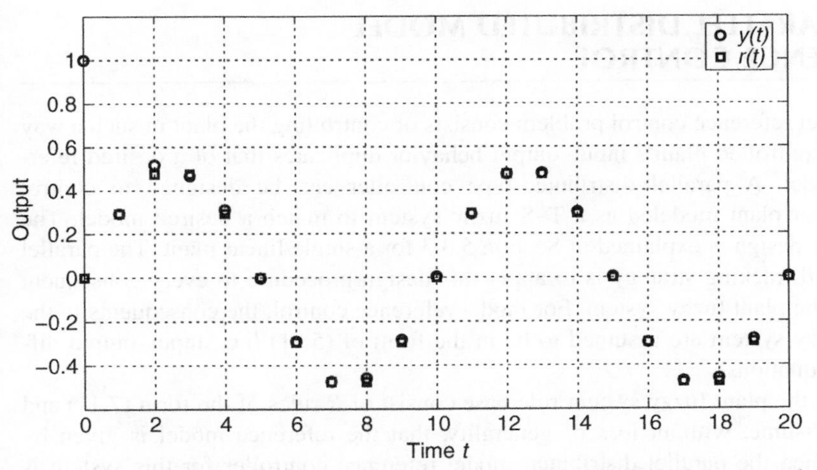

Figure 7.5. Tracking performance $y(t)$ and $r(t)$ of parallel distributed one-step-ahead tracking controller.

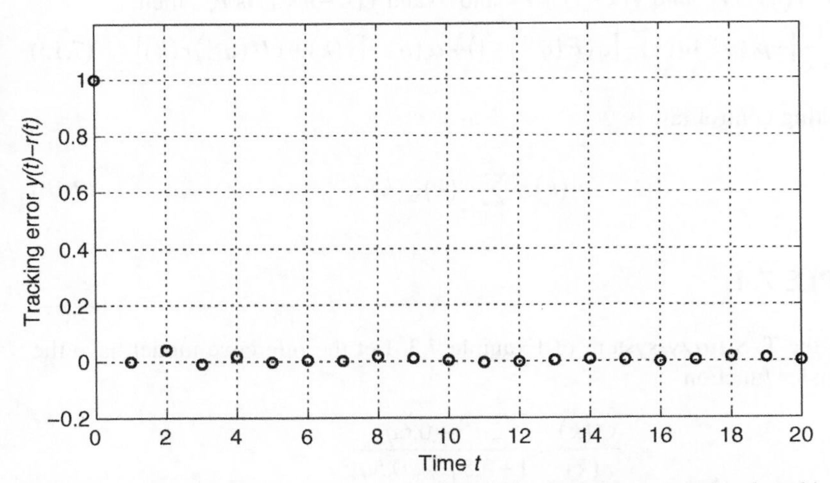

Figure 7.6. Tracking error $y(t) - r(t)$ of parallel distributed one-step-ahead tracking controller.

The tracking is not perfect; for this example the tracking error always remains in the range of $\pm 10^{-3}$ because of the interpolation properties of the T–S fuzzy plant and controller. The tracking would be perfect if each individual control law were applied to each individual plant. This is not the case, however, with parallel distributed-type controllers. Instead, we only have a weighted average of controllers controlling a weighted average of plants. Therefore, the overall control is only an approximation of the controller needed to effect perfect tracking for the overall plant, which is in fact nonlinear.

The Matlab code that produced Figures 7.5 and 7.6 is given in the Appendix.

7.4 PARALLEL DISTRIBUTED MODEL REFERENCE CONTROL

The model reference control problem consists of controlling the plant in such a way that the controlled plant's input–output behavior duplicates that of a desired reference model. A parallel distributed-type controller can be designed to control a nonlinear plant modeled as a T–S fuzzy system to match a desired model. The controller design is explained in Section 5.3.3 for a single linear plant. The parallel distributed tracking strategy is to apply this design procedure to every consequent plant in the plant fuzzy system. For model reference control, the consequents in the plant fuzzy system are assumed to be in the form of (5.11) (i.e., input–output difference equations).

Let the plant fuzzy system rule base consist of R rules of the form (7.11) and (7.12). Assume, without loss of generality, that the reference model is given by (5.34). Then the parallel distributed model reference controller for this system is another fuzzy system with R rules of the form:

R_i: If $y(k)$ is P_1^K and $y(k-1)$ is P_2^L and $\cdots$ and $y(k-n+1)$ is P_n^M, then

$$u^i(k)=\frac{1}{b_{i,0}}\left\{-\beta_i'(q^{-1})u(k)-\left[q\left(E(q^{-1})-1\right)+\alpha_i(q^{-1})\right]y(k)+gH(q^{-1})r(k)\right\} \quad (7.15)$$

The resulting control law is

$$u(k)=\sum_{i=1}^{R}u^i(k)\xi_i(k) \quad (7.16)$$

EXAMPLE 7.4

Consider the T–S fuzzy system of Example 7.3. Let the reference model have the pulse transfer function

$$\frac{y^*(k)}{r(k)}=\frac{2q^{-1}+0.6q^{-2}}{1+0.5q^{-1}+0.5q^{-2}}$$

This can be expressed as (5.34) with $g=2$, $H(q^{-1}) = 1+ 0.3q^{-1}$, and $E(q^{-1}) =1 + 0.5q^{-1} + 0.5q^{-2}$.

As an example, consider again the third consequent plant above, that is,

$$y^3(k+1)=1.2y(k)+0.5y(k-1)+1.5u(k)+0.7u(k-1)$$

For this plant, we have $\alpha_3(q^{-1}) = 1.2 + 0.5q^{-1}$, $\beta_3(q^{-1}) = 1.5 + 0.7q^{-1}$, $b_{3,0} = 1.5$, and $\beta_3'(q^{-1})=0.7q^{-1}$. Therefore, (5.37) yields

$$\gamma(q^{-1})y(k)+(1.5+0.7q^{-1})u(k)=2(1+0.3q^{-1})r(k)$$

where $\gamma(q^{-1})$ is the unique polynomial of order 1 satisfying

$$1+0.5q^{-1}+0.5q^{-2}=1+q^{-1}\left[\gamma(q^{-1})-(1.2+0.5q^{-1})\right]$$

It is straightforward to show that $\gamma(q^{-1}) = \gamma_0 + \gamma_1 q^{-1} = 1.7 + q^{-1}$. Therefore, (5.37) yields the control law

$$u^3(k) = \frac{1}{1.5}[-0.7u(k-1) - 1.7y(k) - y(k-1) + 2r(k) + 0.6r(k-1)]$$

Using the model reference control law (5.37) on each consequent plant yields the parallel distributed model reference controller:

1. If $y(k)$ is P_1^1 and $y(k-1)$ is P_2^1, then

$$u^1(k) = -0.6u(k-1) - 2y(k) - 0.1y(k-1) + 2r(k) + 0.6r(k-1)$$

2. If $y(k)$ is P_1^1 and $y(k-1)$ is P_2^2, then

$$u^2(k) = \frac{1}{1.2}[-u(k-1) - 0.9y(k) + 1.3y(k-1) + 2r(k) + 0.6r(k-1)]$$

3. If $y(k)$ is P_1^2 and $y(k-1)$ is P_2^1, then

$$u^3(k) = \frac{1}{1.5}[-0.7u(k-1) - 1.7y(k) - y(k-1) + 2r(k) + 0.6r(k-1)]$$

4. If $y(k)$ is P_1^2 and $y(k-1)$ is P_2^2, then

$$u^4(k) = \frac{1}{1.5}[-u(k-1) - 1.3y(k) + 1.1y(k-1) + 2r(k) + 0.6r(k-1)]$$

Let the input fuzzy sets P_1^1, P_1^2, P_2^1, and P_2^2 be defined as in Figure 6.4. Let the desired reference signal be $r(k) = 0.5\sin(0.2\pi k)$. With $\Delta t = 1$ s (hence, $t = k\Delta t = k$), initial conditions $y(0) = 1$, $y(-1) = -1$, a plot of the output $y(t)$ of the controlled plant together with the model output $y_m(t)$ is shown in Figure 7.7 and the corresponding error $y(t) - y_m(t)$ is shown in Figure 7.8.

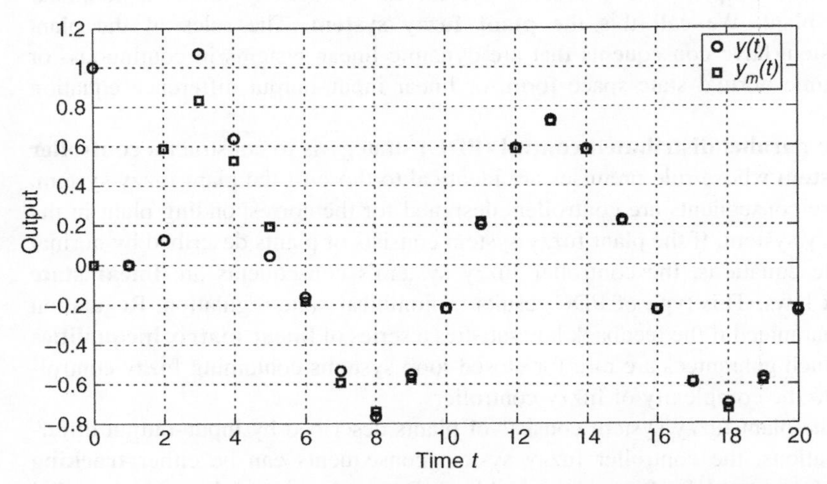

Figure 7.7. Model matching performance $y(t)$ and $y_m(t)$ of parallel distributed model reference controller.

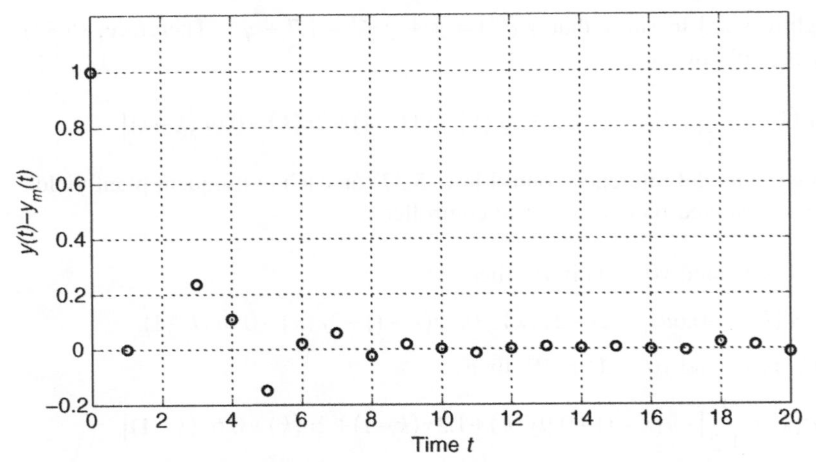

Figure 7.8. Model matching performance $y(t) - y_m(t)$ of parallel distributed model reference controller.

The model following is not perfect; the absolute value of the error between $y(t)$ and $y_m(t)$ for this example always remains in the range of 0.025 because of the interpolation properties of the T–S fuzzy plant and controller.

The Matlab code that produced Figures 7.7 and 7.8 is given in the Appendix.

7.5 SUMMARY

As shown in Chapter 6, a T–S fuzzy system can be used to model a nonlinear dynamic plant. We call this the **plant fuzzy system**. The rules of the plant fuzzy system have consequents that are dynamic linear systems in continuous- or discrete-time normal state space form, or linear input–output difference equation form.

The **parallel distributed control (PDC)** strategy is to construct a **controller fuzzy system** whose rule premises are identical to those of the plant fuzzy system, and whose consequents are controllers designed for the corresponding plant in the plant fuzzy system. If the plant fuzzy system consists of plants described by normal form state equations, the controller fuzzy system's consequents are **linear state feedback** laws. This type of PDC results in nonlinear state regulation. Regulation can be guaranteed if the feedback law satisfies a series of **linear matrix inequalities (LMI)**. Such guarantees are rare for closed-loop systems containing fuzzy controllers due to the complexity of fuzzy controllers.

If the plant fuzzy system consists of plants described by input–output difference equations, the controller fuzzy system consequents can be either **tracking control** laws or **model reference control** laws. Remember that in the case of parallel distributed output tracking or parallel distributed model reference control, there is no formal proof that the control objective (i.e., output tracking or model following) will be achieved. Nevertheless, these two approaches will be shown in Chapter 10 to be very useful for adaptive fuzzy control.

EXERCISES

7.1 Consider the system of Example 7.1. Design a parallel distributed controller that places both closed-loop poles at −1.

7.2 Prove that the controller designed in Problem 7.1 stabilizes the closed-loop system.

7.3 Simulate the closed-loop system designed in Problem 7.1. Use fourth-order Runge–Kutta integration with a step size of 0.01 s. Let $x_1(0) = 1$, $x_2(0) = -1$, with fuzzy sets P^1 and P^2 specified by two triangular membership functions with midpoints at $x = -2, 2$ such that they form a partition of unity.

7.4 Consider the system of Example 7.2. Design a parallel distributed controller that places the closed-loop poles at $0.9 \pm j0.1$.

7.5 Prove that the controller designed in Problem 7.4 stabilizes the closed-loop system.

7.6 Simulate the closed-loop system designed in Problem 7.4. Let $x_1(0) = 1$, $x_2(0) = -1$, with fuzzy sets P^1 and P^2 specified by two triangular membership functions with midpoints at $x = -2, 2$ such that they form a partition of unity.

7.7 Consider the following T–S system:

1. If $y(k)$ is P_1^1 and $y(k-1)$ is P_2^1, then

$$y^1(k+1) = y(k) - y(k-1) + 0.2u(k) + 0.1u(k-1)$$

2. If $y(k)$ is P_1^1 and $y(k-1)$ is P_2^2, then

$$y^2(k+1) = 1.2y(k) + 0.5y(k-1) + 1.5u(k) + 0.7u(k-1)$$

3. If $y(k)$ is P_1^2 and $y(k-1)$ is P_2^1, then

$$y^3(k+1) = 0.4y(k) - 1.8y(k-1) + 1.2u(k) + u(k-1)$$

4. If $y(k)$ is P_1^2 and $y(k-1)$ is P_2^2, then

$$y^4(k+1) = 0.3y(k) - 0.6y(k-1) + 1.5u(k) + u(k-1)$$

Design a parallel distributed tracking controller to make this system track the reference signal $r(k)$.

7.8 Simulate the closed-loop system designed in Problem 7.7. Let $x_1(0) = 1$, $x_2(0) = -1$, with fuzzy sets P^1 and P^2 specified by two Gaussian membership functions with midpoints at $x = -2, 2$ such that they cross at 0.4. Let $r(k) = 0.1(1 - \cos(0.3k))$. Plot $y(k) - r(k)$, where $y(k)$ is the output of the controlled plant.

7.9 Consider the system of Problem 7.7. Design a parallel distributed model reference controller to make the controlled system's model match the model whose pulse transfer function is

$$G\left(q^{-1}\right)=\frac{q^{-1}+0.2q^{-2}}{1+0.6q^{-1}+0.7q^{-2}}$$

7.10 Simulate the closed-loop system designed in Problem 7.9. Let $x_1(0) = 1$, $x_2(0) = -1$, with fuzzy sets P^1 and P^2 specified by two triangular membership functions with midpoints at $x = -2, 2$, such that they form a partition of unity. Let the input to the reference model be $r(k) = 0.1(1 - \cos(0.3k))$. Plot $y(k) - y_m(k)$, where $y(k)$ is the output of the controlled plant and $y_m(k)$ is the output of the reference model.

CHAPTER 8

ESTIMATION OF STATIC NONLINEAR FUNCTIONS FROM DATA

In Chapters 3 and 6, we saw that some fuzzy systems have nonlinear memoryless input–output characteristic functions. The nonlinear characteristic function of these fuzzy systems is a result of the rules, T-norm, membership functions, and defuzzification method chosen by the designer. In these systems, the input and output membership functions are fixed *a priori*.

It is also possible, given a particular nonlinear function, to find a fuzzy system whose input–output characteristic matches it. In the fuzzy system, the output and in some cases input membership functions are not fixed *a priori*, but are adjusted so that the input–output characteristic most closely matches the nonlinear function in some sense. The determination of a fuzzy system to approximate a given nonlinear function is done utilizing well-known results from estimation theory. The estimation methods introduced in this chapter are the least-squares and gradient approaches. These are introduced because they have direct applications to control.

It will be seen in Chapter 9 that the ability to model static nonlinear functions as fuzzy systems makes it possible to model dynamic nonlinear systems as fuzzy systems. The motivation for modeling a dynamic nonlinear system as a fuzzy system is that with such a model, the parallel distributed control methods of the previous chapter can be employed to control the system. This enables a very powerful type of control known as *adaptive fuzzy control*, introduced in Chapter 10. With adaptive fuzzy control, tracking and model reference controllers can be designed for unknown nonlinear systems.

8.1 LEAST-SQUARES ESTIMATION

The basic idea in least-squares estimation is to find parameters that minimize the square of the difference between the estimated and true nonlinear functions. We first introduce the *batch least-squares* approach, which finds the minimum based on a finite number of stored input–output measurements from the plant. This process is then made recursive resulting in an algorithm, called *recursive least squares*, that updates the parameter estimate at each time step as new measurements are obtained from the plant.

8.1.1 Batch Least Squares [4,32]

Consider the algebraic system:

$$y(j)=\phi_1(j)\theta_1+\phi_2(j)\theta_2+\cdots+\phi_R(j)\theta_R \tag{8.1}$$

where $y(j)$, $\phi_i(j)$, $i = 1, 2, \ldots, R, j = 1, 2, 3, \ldots$ are known and $\theta_1, \ldots, \theta_R$ are unknown constant parameters to be determined.

Defining the vectors

$$\begin{aligned}\phi(j)&=[\phi_1(j)\quad \phi_2(j)\quad \cdots\quad \phi_R(j)]^T \sim R\times 1\\ \theta&=[\theta_1\quad \theta_2\quad \cdots\quad \theta_R]^T \sim R\times 1\end{aligned}$$

the above system can be written as

$$y(j)=\phi^T(j)\theta \tag{8.2}$$

This is called a *regression model* and $\phi(j)$ is called the *regressor*. Note that $y(j)$ is *linear in the parameter vector* θ.

Let k pairs of observations and corresponding regressors

$$\{y(j),\phi(j),\ j=1,2,\cdots,k\}$$

be obtained from k experiments performed on the system. This is sometimes called *training data*. The objective is to estimate the unknown parameter vector θ from the training data. The estimate will be called $\hat{\theta}$.

To find the best $\hat{\theta}$, consider the performance measure

$$V_1(\hat{\theta})=\frac{1}{2}\sum_{j=1}^{k}\left[y(j)-\phi^T(j)\hat{\theta}\right]^2 \tag{8.3}$$

Then the best $\hat{\theta}$, in a least-squares sense, is the one that minimizes $V_1(\hat{\theta})$. Defining

$$Y(k)=[y(1)\quad y(2)\quad \cdots\quad y(k)]^T \sim k\times 1$$

$$\Phi(k)=\begin{bmatrix}\phi^T(1)\\ \phi^T(2)\\ \vdots\\ \phi^T(k)\end{bmatrix}\sim k\times R \tag{8.4}$$

then we can write the performance measure (8.3) as

$$\begin{aligned}V_1(\hat{\theta}(k))&=\frac{1}{2}\left(Y(k)-\Phi(k)\hat{\theta}(k)\right)^T\left(Y(k)-\Phi(k)\hat{\theta}(k)\right)\\ &=\frac{1}{2}\left(Y^TY-Y^T\Phi\hat{\theta}-\hat{\theta}^T\Phi^TY+\hat{\theta}^T\Phi^T\Phi\hat{\theta}\right)\\ &=\frac{1}{2}\left(Y^TY-2\hat{\theta}^T\Phi^TY+\hat{\theta}^T\Phi^T\Phi\hat{\theta}\right)\end{aligned} \tag{8.5}$$

where the dependence on k has been dropped for notational simplicity. The performance measure $V_1(\hat{\theta})$ is quadratic and convex (i.e., bowl shaped with one unique minimum) in parameter space. Therefore, to minimize $V_1(\hat{\theta})$ set

$$\frac{\partial V_1(\hat{\theta})}{\partial \hat{\theta}} = 0$$

and solve for $\hat{\theta}$:

$$\frac{\partial V_1(\hat{\theta})}{\partial \hat{\theta}} = -\Phi^T Y + \Phi^T \Phi \hat{\theta} = 0$$

Therefore, $\hat{\theta}$ satisfying

$$\Phi^T \Phi \hat{\theta} = \Phi^T Y$$

minimizes $V_1(\hat{\theta})$. If the matrix $\Phi^T \Phi$ is invertible, the minimum is unique and is given by

$$\hat{\theta}^*(k) = (\Phi^T \Phi)^{-1} \Phi^T Y \tag{8.6}$$

This is the *best least-squares estimate of* θ based on the k measurements. It is a *batch* calculation because we have to wait until k observations are made before we can obtain $\hat{\theta}^*$.

The fact that $\hat{\theta}^*$ is *unique* is a result of $V_1(\hat{\theta})$ in (8.5) being quadratic and convex, which is a result of y in (8.2) being *linear in the parameter estimates* $\hat{\theta}$. The requirement that $\Phi^T\Phi$ be nonsingular is a *persistent excitation* requirement on the regressor, which often consists of present and past values of the input or a combination of the input and output or functions of these. Persistent excitation of the regressor means that the regressor contains enough variation to permit accurate, unique determination of $\hat{\theta}^*$. It is always a requirement on the regressor frequency content in all types of identification and estimation methods, fuzzy or not.

8.1.2 Recursive Least Squares

It is often desirable to make the estimation process recursive rather than batch. This way, we would have an estimate $\hat{\theta}$ at every time step rather than having to wait k time steps for one. When the batch least-squares estimation method is made recursive, the parameter estimate at time k, $\hat{\theta}(k)$, is expressed in terms of the previous estimate, $\hat{\theta}(k-1)$ as follows.

Theorem: Let a single-output algebraic system be described by $y(k) = \phi^T(x(k))\theta$, where $x(k)$ is the known system input at time k, $y(k)$ is the known system output at time k, $\phi(x(k)) \sim R \times 1$ is a vector of known functions of the input, and $\theta \sim R \times 1$ is an unknown vector of constants. Given initial conditions $\hat{\theta}(k_0)$ and $P(k_0) \sim R \times R$ symmetric positive definite, the recursive least-squares estimate $\hat{\theta}(k)$ satisfies:

$$\hat{\theta}(k) = \hat{\theta}(k-1) + K(k)\left[y(k) - \phi^T(k)\hat{\theta}(k-1)\right] \tag{8.7a}$$

$$K(k) = P(k-1)\phi(k)\left[I + \phi^T(k)P(k-1)\phi(k)\right]^{-1} \tag{8.7b}$$

$$P(k) = \left[I - K(k)\phi^T(k)\right]P(k-1) \tag{8.7c}$$

Notice that for a single-output system, the bracketed term in (8.7b) is scalar. Hence, in this case no matrix inversion is required, significantly reducing the computational load. In (8.7a), the new parameter estimate $\hat{\theta}(k)$ is equal to the current one $\hat{\theta}(k-1)$ plus a correction term that is a measure of how well the current estimated system approximates the true process. In the term $\left[y(k)-\phi^T(k)\hat{\theta}(k-1)\right]$, $y(k)$ is the output of the true process while $\phi^T(k)\hat{\theta}(k-1)$ is the output of the current estimated system.

Proof: This can be derived from the batch least-squares estimate (8.6) by writing $\hat{\theta}^*(k)$ in terms of $\hat{\theta}^*(k-1)$, defining $P(k) = [\Phi^T(k)\Phi(k)]^{-1}$, and using the Matrix Inversion Lemma: $[A + BCD]^{-1} = A^{-1} - A^{-1}B(DA^{-1}B + C^{-1})^{-1}DA^{-1}$ [33].

8.2 BATCH LEAST-SQUARES FUZZY ESTIMATION IN MAMDANI FORM

Assume a fuzzy system with n inputs $x = [x_1, x_2, \ldots, x_n]$, a single output y, and R rules in the form

$$R_i: \quad \text{If } x_1 \text{ is } A_1^K \text{ and } x_2 \text{ is } A_2^L \text{ and } \cdots \text{ and } x_n \text{ is } A_n^M, \text{ then } y \text{ is } B^i \tag{8.8}$$

Each rule has its own unique consequent fuzzy set B^i characterized by a singleton membership function located at $y = b^i$, $i = 1, \ldots, R$, which is to be determined.

If the ith rule's degree of firing (i.e., premise membership function) is $\mu_i(x)$, the crisp output of the system is

$$\begin{aligned} f(x) &= \frac{\sum_{i=1}^{R} b^i \mu_i(x)}{\sum_{i=1}^{R} \mu_i(x)} \\ &= b^1\xi_1(x) + b^2\xi_2(x) + \cdots + b^R\xi_R(x) \end{aligned} \tag{8.9}$$

where $\xi_i(x)$ is the fuzzy basis function for the ith rule:

$$\xi_i(x) = \frac{\mu_i(x)}{\sum_{i=1}^{R} \mu_i(x)} \tag{8.10}$$

Since the input membership functions are fixed, the premise membership functions $\mu_i(x)$ and the fuzzy basis functions $\xi_i(x)$ are known *a priori*. Since the output membership function locations b^i are unknown, the fuzzy system (8.9) is in the form of (8.2), with $\phi^T(x) = [\xi_1(x) \;\; \xi_2(x) \;\; \cdots \;\; \xi_R(x)]$ and $\theta = \left[b^1 \;\; b^2 \;\; \cdots \;\; b^R\right]^T$. Therefore, batch or recursive least squares can be used to estimate the θ resulting in the fuzzy system $f(x)$ that best approximates a function $g(x)$ in a least-squares sense.

EXAMPLE 8.1 (SINGLE INPUT FUNCTION)

Consider the function

$$g(x) = x - \cos(1.5x) + \sin(0.4x) \tag{8.11}$$

The output g is a nonlinear function of the input x. We wish to find a fuzzy system $f(x|\theta)$ that approximates the nonlinear function $g(x)$ over the interval $0 \le x \le 6$. Assume that we do this with a four-rule adjustable Mamdani fuzzy system. Let the input fuzzy sets be characterized by Gaussian membership functions $\mu_i(x) = \exp(-0.5((x - c^i)/\sigma^i)^2)$, $i = 1, 2, 3, 4$ with centers $c^1 = 0$, $c^2 = 2$, $c^3 = 4$, $c^4 = 6$ and spreads $\sigma^1 = \sigma^2 = \sigma^3 = \sigma^4 = 1.184$. The centers are chosen to cover the input universe of discourse [0 6] evenly. The spreads are chosen so that adjacent memberships cross at 0.7. The input memberships are shown in Figure 8.1.

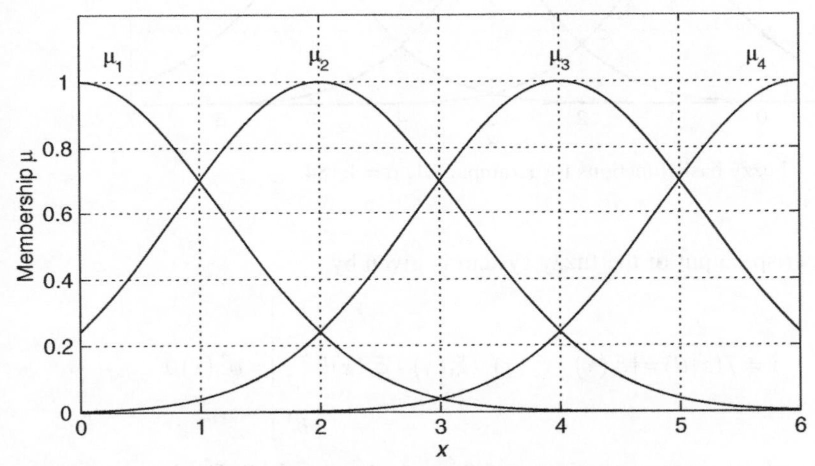

Figure 8.1. Input membership functions for Example 8.1, $\sigma = 1.184$.

Using *product* T-norm, the fuzzy basis functions are

$$\xi_j(x) = \frac{\exp\left(-\frac{1}{2}\left(\frac{x-c^j}{\sigma^j}\right)^2\right)}{\sum_{i=1}^{4}\exp\left(-\frac{1}{2}\left(\frac{x-c^i}{\sigma^i}\right)^2\right)} \tag{8.12}$$

for $j = 1, 2, 3, 4$. These are plotted in Figure 8.2.

Since there are four rules, there are four unknown parameters (i.e., the output centers $b^1, \ldots, b^4$). We will estimate these with batch least squares. The input portion of the training data should sufficiently cover the domain over which we desire a good estimate of g. Since this domain is $0 \le x \le 6$, let the input portion of the training data be the seven numbers $x = [0 \quad 1 \quad 2 \quad 3 \quad 4 \quad 5 \quad 6]$. The output portion of the training data consists of the seven corresponding values of g, that is, $g = [-1 \quad 1.3187 \quad 3.7073 \quad 4.1428 \quad 4.0394 \quad 5.5627 \quad 7.5866]$. Thus, the training data in input–output pair form is

$$\{(0,-1),(1,1.3187),(2,3.7073),(3,4.1428),(4,4.0394),(5,5.5627),(6,7.5866)\}$$

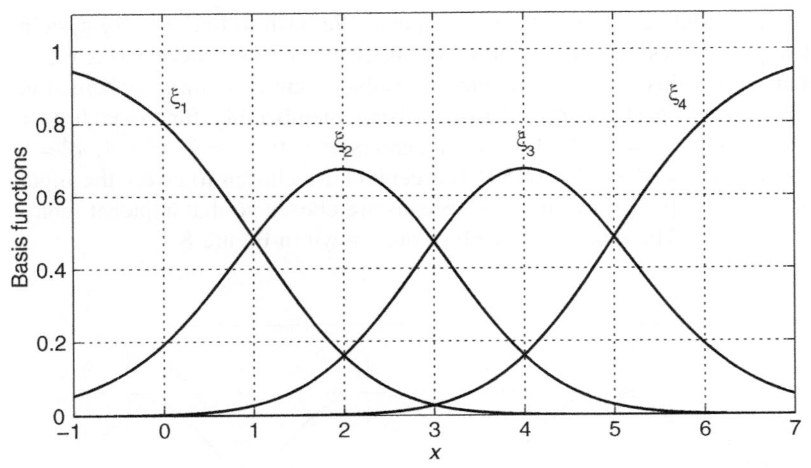

Figure 8.2. Fuzzy basis functions for Example 8.1, $\sigma = 1.184$.

The crisp output of the fuzzy system is given by

$$y = f(x|\theta) = [\xi_1(x) \quad \xi_2(x) \quad \xi_3(x) \quad \xi_4(x)] \begin{bmatrix} b^1 \\ b^2 \\ b^3 \\ b^4 \end{bmatrix} = \phi^T(x)\theta$$

Then the batch least-squares estimate of θ is calculated as in (8.6) where

$$Y = [-1 \quad 1.3187 \quad 3.7073 \quad 4.1428 \quad 4.0394 \quad 5.5627 \quad 7.5866]^T$$

and

$$\Phi = \begin{bmatrix} \xi_1(x(1)) & \xi_2(x(1)) & \xi_3(x(1)) & \xi_4(x(1)) \\ \xi_1(x(2)) & \xi_2(x(2)) & \xi_3(x(2)) & \xi_4(x(2)) \\ \xi_1(x(3)) & \xi_2(x(3)) & \xi_3(x(3)) & \xi_4(x(3)) \\ \xi_1(x(4)) & \xi_2(x(4)) & \xi_3(x(4)) & \xi_4(x(4)) \\ \xi_1(x(5)) & \xi_2(x(5)) & \xi_3(x(5)) & \xi_4(x(5)) \\ \xi_1(x(6)) & \xi_2(x(6)) & \xi_3(x(6)) & \xi_4(x(6)) \\ \xi_1(x(7)) & \xi_2(x(7)) & \xi_3(x(7)) & \xi_4(x(7)) \end{bmatrix}$$

$$= \begin{bmatrix} 0.8042 & 0.1931 & 0.0027 & 0.0000 \\ 0.4859 & 0.4859 & 0.0280 & 0.0001 \\ 0.1618 & 0.6741 & 0.1618 & 0.0022 \\ 0.0273 & 0.4727 & 0.4727 & 0.0273 \\ 0.0022 & 0.1618 & 0.6741 & 0.1618 \\ 0.0001 & 0.0280 & 0.4859 & 0.4859 \\ 0.0000 & 0.0027 & 0.1931 & 0.8042 \end{bmatrix}$$

Note that it is necessary to evaluate the fuzzy basis functions at the inputs of the data set, that is, at $x(1)$, $x(2)$, … , $x(7)$. Therefore, for example,

$$\xi_2(x(4)) = \frac{\exp\left(-\frac{1}{2}\left(\frac{3-2}{1.184}\right)^2\right)}{\Delta} = 0.4727$$

where

$$\Delta = \exp\left(-\frac{1}{2}\left(\frac{3-0}{1.184}\right)^2\right) + \exp\left(-\frac{1}{2}\left(\frac{3-2}{1.184}\right)^2\right) + \exp\left(-\frac{1}{2}\left(\frac{3-4}{1.184}\right)^2\right) + \exp\left(-\frac{1}{2}\left(\frac{3-6}{1.184}\right)^2\right)$$
$$= 0.0404 + 0.7 + 0.7 + 0.0404$$

The batch LS algorithm (8.6) gives the following parameters:

$$\hat{\theta}^* = \begin{bmatrix} b^1 \\ b^2 \\ b^3 \\ b^4 \end{bmatrix} = \begin{bmatrix} -2.6660 \\ 5.4934 \\ 2.6386 \\ 8.7027 \end{bmatrix}$$

This results in the fuzzy characteristic $f(x|\hat{\theta}^*)$ shown in Figure 8.3. Even with so few fuzzy sets and data points, the fit is still quite good.

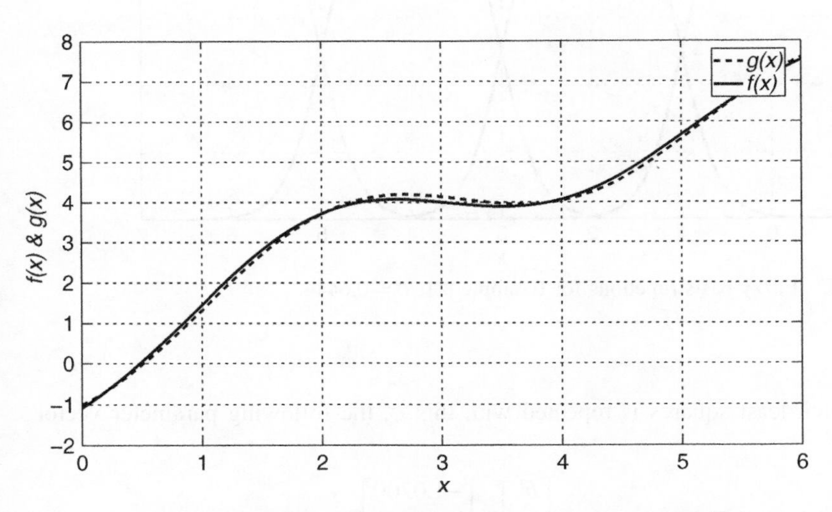

Figure 8.3. Nonlinear function $g(x)$ and its fuzzy approximation $f(x|\hat{\theta}^*)$, $\sigma = 1.184$.

To see the effect of varying the spreads of the input membership functions, let us keep the same centers with smaller spreads, that is, $\sigma = 0.6444$, which results in adjacent Gaussians crossing at 0.3 (see Figs. 8.4 and 8.5).

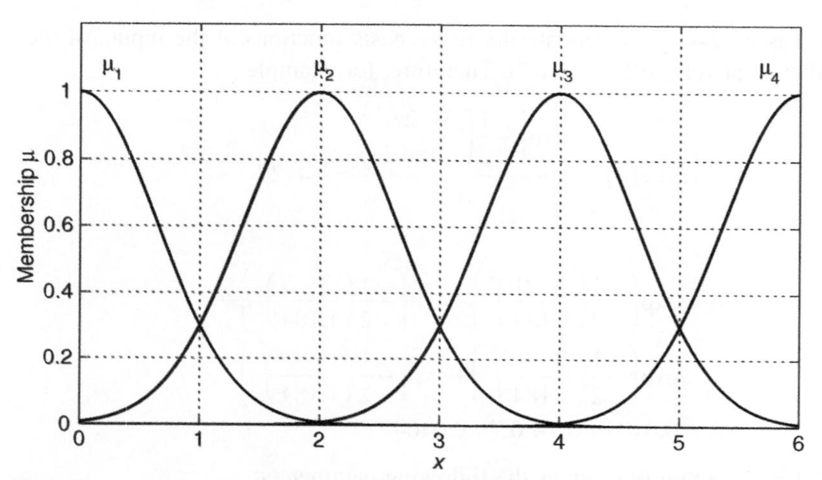

Figure 8.4. Input membership functions for Example 8.1, $\sigma = 0.6444$.

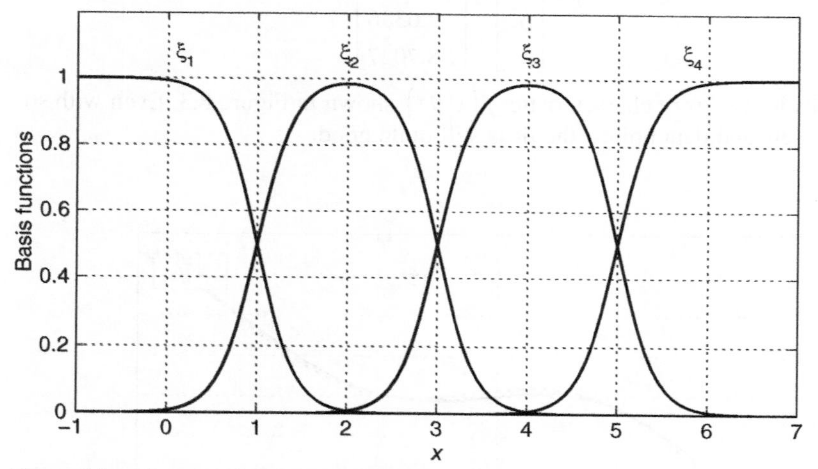

Figure 8.5. Fuzzy basis functions for Example 8.1, $\sigma = 0.6444$.

When batch least squares is repeated with this σ, the following parameter vector results:

$$\hat{\theta}^* = \begin{bmatrix} b^1 \\ b^2 \\ b^3 \\ b^4 \end{bmatrix} = \begin{bmatrix} -1.0700 \\ 3.8262 \\ 4.0212 \\ 7.5119 \end{bmatrix}$$

The fuzzy characteristic is shown in Figure 8.6.

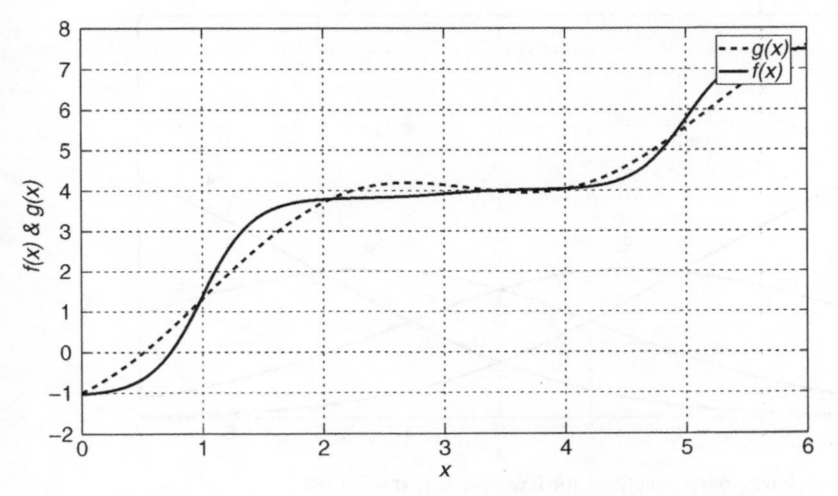

Figure 8.6. Nonlinear function $g(x)$ and its fuzzy approximation $f\left(x\middle|\hat{\theta}*\right)$, $\sigma = 0.6444$.

The fit is not so good with $\sigma = 0.6444$. The input Gaussians (Fig. 8.4), and hence the basis functions (Fig. 8.5), vary too quickly to accurately approximate the more gently sloping function $g(x)$.

If we try larger spreads, that is, $\sigma = 2.1784$, this results in adjacent Gaussians crossing at 0.9 (see Figs. 8.7 and 8.8).

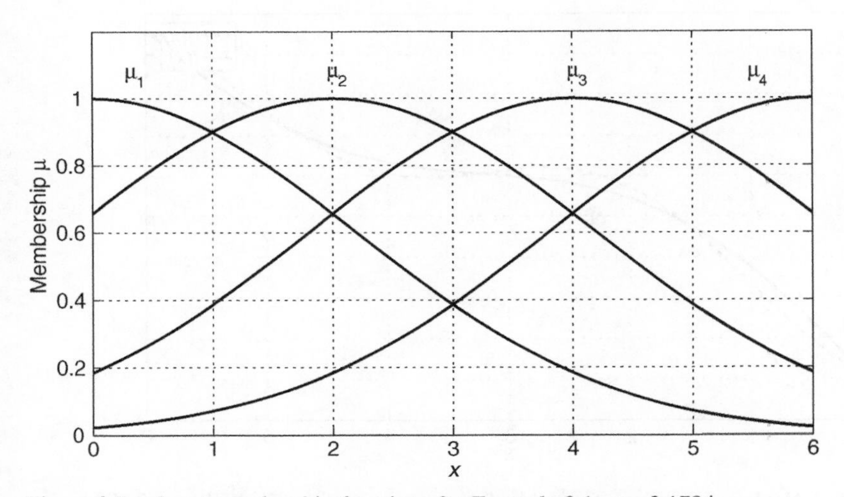

Figure 8.7. Input membership functions for Example 8.1, $\sigma = 2.1784$.

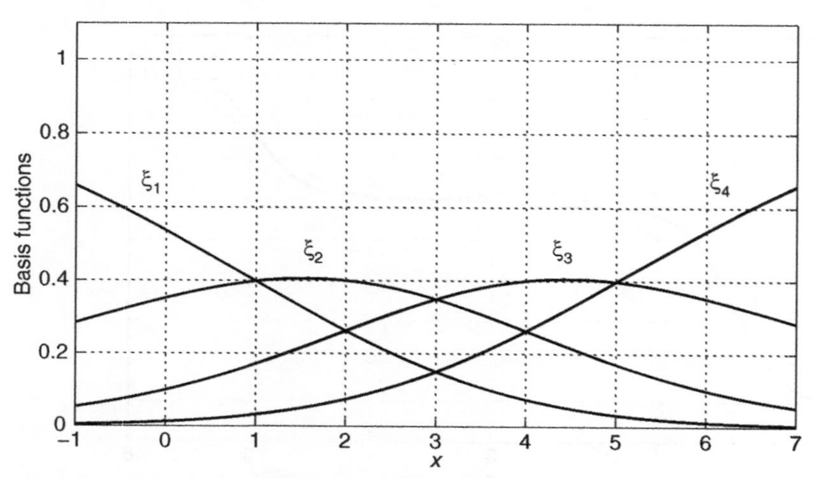

Figure 8.8. Fuzzy basis functions for Example 8.1, $\sigma = 2.1784$.

When batch least squares is repeated with this σ, the following parameter vector results:

$$\hat{\theta}^* = \begin{bmatrix} b^1 \\ b^2 \\ b^3 \\ b^4 \end{bmatrix} = \begin{bmatrix} -18.0655 \\ 28.8548 \\ -18.9838 \\ 21.6675 \end{bmatrix}$$

The resulting fuzzy characteristic is shown in Figure 8.9.

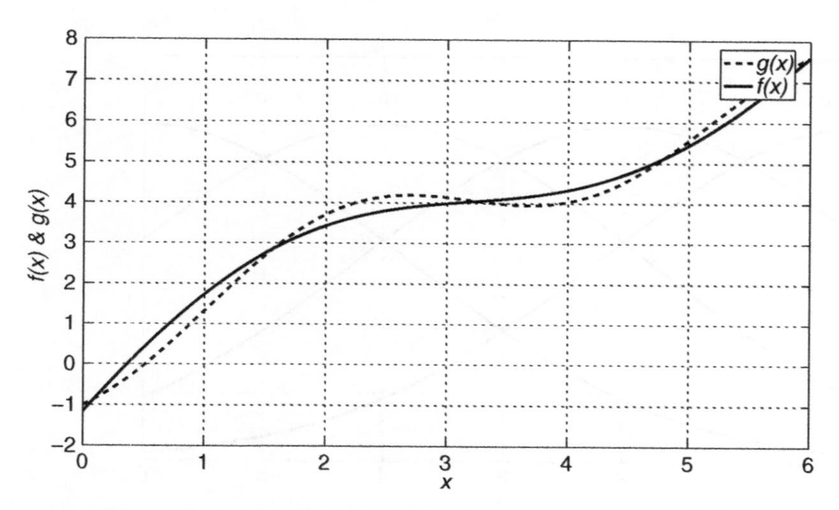

Figure 8.9. Nonlinear function $g(x)$ and its fuzzy approximation $f\left(x\middle|\hat{\theta}^*\right)$, $\sigma = 2.1784$.

Again, the fit is not so good with $\sigma = 2.1784$. The input Gaussians (Fig. 8.7), and hence the basis functions (Fig. 8.8), vary too slowly to accurately approximate the more quickly changing function $g(x)$.

This example illustrates the importance of choosing good input membership functions when using least squares. The Matlab code producing the results for this example is found in the Appendix .

EXAMPLE 8.2 (MULTIPLE INPUT FUNCTION)

Consider the function

$$g(x_1, x_2) = \sin x_1 \cos^2 x_2 \tag{8.13}$$

The output g is a nonlinear function of the inputs x_1 and x_2, shown in Figure 8.10.

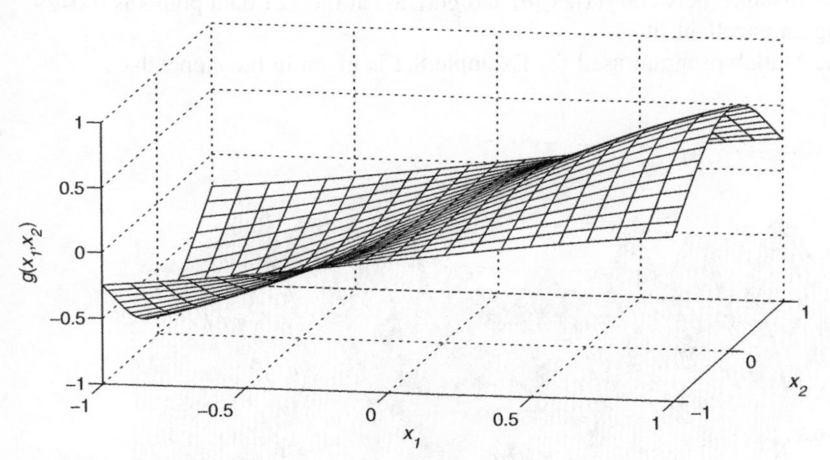

Figure 8.10. The function $g(x_1, x_2) = \sin x_1 \cos^2 x_2$.

We wish to find a fuzzy system $f((x_1, x_2)|\theta)$ that approximates the nonlinear function $g(x_1, x_2)$ over the domain $-1 \le x_1 \le 1$, $-1 \le x_2 \le 1$. Let us do this with an adjustable Mamdani fuzzy system with two inputs x_1 and x_2. On each universe, let there be five Gaussian membership functions with centers $[-1 \quad -0.5 \quad 0 \quad 0.5 \quad 1]$ and spreads $\sigma = 0.2123$ (so adjacent memberships cross at 0.5). For the rule premises, use all possible combinations of these five fuzzy sets on both universes resulting in 25 rules, each with its own adjustable consequent singleton fuzzy set b^i, $i = 1, 2, \ldots, 25$.

The fuzzy basis function for rule i is

$$\xi_i(x_1, x_2) = \frac{\mu_i(x_1, x_2)}{\sum_{j=1}^{25} \mu_j(x_1, x_2)}$$

where $\mu_i(x_1, x_2)$ is the premise value (i.e., degree of firing) of the ith rule. The crisp output of this system is

$$f(x_1, x_2 | \theta) = \sum_{i=1}^{25} \xi_i(x_1, x_2) b^i = \phi^T \theta$$

The input portion of the training data should sufficiently cover the domain over which we desire a good estimate of g. Since this domain is $-1 \le x_1 \le 1$, $-1 \le x_2 \le 1$, let the input portion of the training data consist of $11^2 = 121$ points uniformly covering this domain, that is, all combinations of $x_1 = [-1, -0.8, -0.6, \ldots, 1]$, $x_2 = [-1, -0.8, -0.6, \ldots, 1]$. The output portion of the training data consists of the 121 corresponding values of $g(x_1, x_2) = \sin x_1 \cos^2 x_2$.

The batch least-squares estimate of θ is calculated as in (8.6), where $\hat{\theta}^* \sim 25 \times 1$, $\Phi \sim 121 \times 25$, and $Y \sim 121 \times 1$. This results in a 2 input, 1 output fuzzy system whose crisp output $f(x_1, x_2)$ closely matches $g(x_1, x_2)$. As a measure of the closeness, the error surface $|f(x_1, x_2) - g(x_1, x_2)|$ is shown in Figure 8.11. The maximum distance between $f(x_1, x_2|\theta)$ and $g(x_1, x_2)$ at the 121 data points is 0.0429, indicating an excellent fit.

The Matlab program used for Example 8.2 is given in the Appendix .

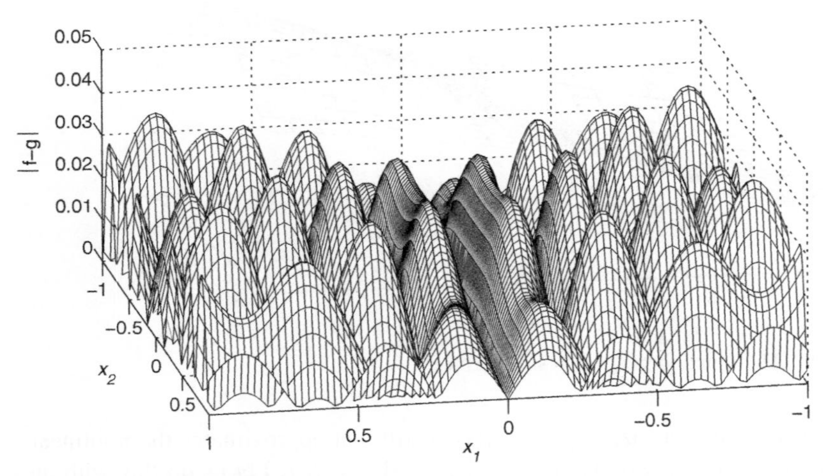

Figure 8.11. Error surface $|f - g|$ for Example 8.2.

8.3 RECURSIVE LEAST-SQUARES FUZZY ESTIMATION IN MAMDANI FORM

Assume a Mamdani fuzzy system with rules in the form of (8.8). The resulting fuzzy system can again be described by (8.9), where the fuzzy basis functions $\xi_i(x)$ are fixed and known. Then the RLS algorithm proceeds as in (8.7) with $\hat{\theta}(k) = \left[\hat{b}^1(k), \hat{b}^2(k), \ldots, \hat{b}^R(k)\right]^T$, $y(k) = g(x(k))$, and $\phi^T(k) = \left[\xi^1(x(k))\xi^2(x(k)) \cdots \xi^R(x(k))\right]$.

EXAMPLE 8.3 (SINGLE INPUT FUNCTION)

Consider again the system of Example 8.1 given by (8.11). This time, we will use RLS to recursively find a fuzzy system $f(x|\theta)$ that approximates the nonlinear function $g(x)$. Again, assume that we do this with an adjustable Mamdani fuzzy system with four rules, four input fuzzy sets characterized by Gaussian membership functions, and four adjustable output fuzzy sets characterized by singleton membership functions. As above, we fix the input membership function centers and spreads. Then the crisp output of the fuzzy system is given by (8.9).

Then, $\hat{\theta}(k)\left[\hat{b}^1(k) \quad \hat{b}^2(k) \quad \hat{b}^3(k) \quad \hat{b}^4(k)\right]$ would be recursively estimated as in (8.7), where $y(k)$ is the measured system output at time k (i.e., $g(x(k))$), and

$$\phi^T(k) = [\xi_1(x(k)) \quad \xi_2(x(k)) \quad \xi_3(x(k)) \quad \xi_4(x(k))]$$

is evaluated at each time k.

The time-varying fuzzy system is

$$f\left(x(k)\middle|\hat{\theta}(k)\right) = \frac{\sum_{i=1}^{4} \hat{b}^i(k)\mu_i(x(k))}{\sum_{i=1}^{4} \mu_i(x(k))}$$

This fuzzy system changes for each k, but converges to the fuzzy system that best approximates $g(x)$ as $k \to \infty$ in a least-squares sense given the chosen input $x(k)$.

The parameter estimates resulting from the RLS algorithm with $x(k) = e(k)$, a random variable uniformly distributed in [0 6] for each k, $\theta(0) = [0 \quad 0 \quad 0 \quad 0]^T$, and $P(0) = 10^5 I$ are shown in Figure 8.12.

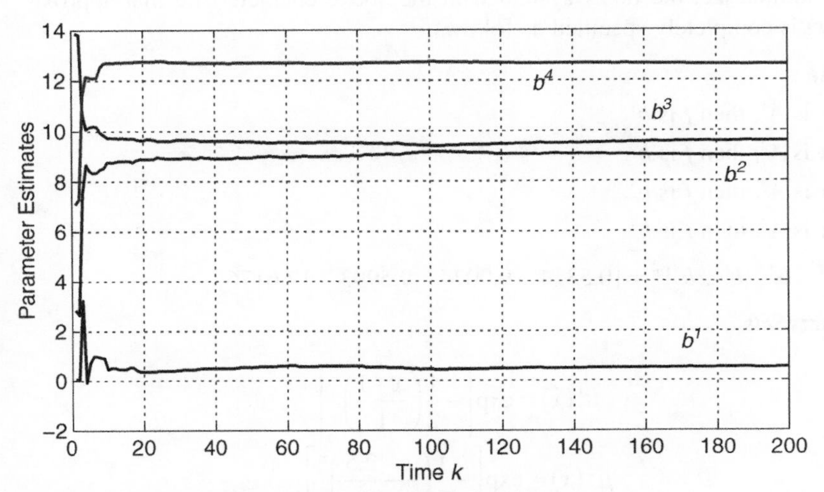

Figure 8.12. Parameter estimates from RLS algorithm.

The parameter estimates start at their initial conditions, transition through time, and converge after ~200 steps to the constant values $[b^1(200) \quad b^2(200) \quad b^3(200) \quad b^4(200)] = [0.5314 \quad 9.0074 \quad 9.5993 \quad 12.6176]$. These final parameter values are not unique, however. They depend on the input $x(k)$, which is a random sequence. Therefore, the final parameter values will be different every time the algorithm is run.

Figure 8.13 shows the fuzzy characteristic compared with $g(x)$ for the above parameter values. The fit is quite good; however it could be better if a more complex fuzzy system (i.e., more rules) were used.

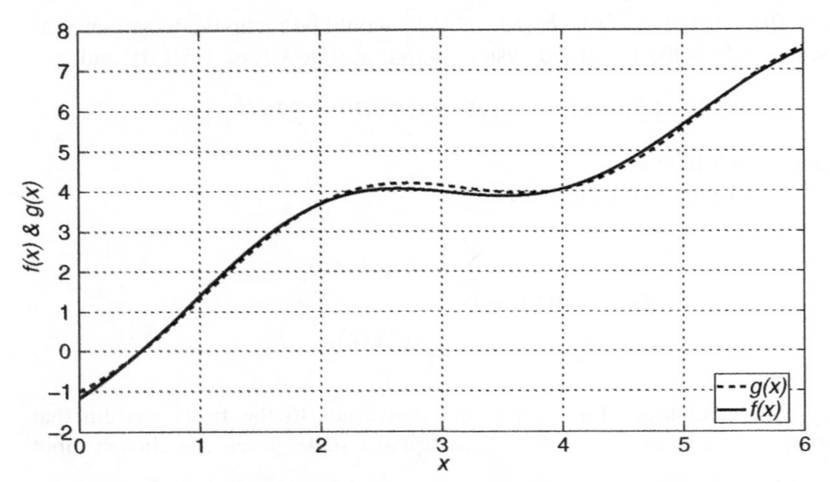

Figure 8.13. Nonlinear function $g(x)$ and its fuzzy approximation $f(x|\theta)$ with $\theta = [b^1, b^2, b^3, b^4]^* = [0.5314 \quad 9.0074 \quad 9.5993 \quad 12.6176]$.

To summarize, the fuzzy system with the above characteristic that approximates $g(x)$ is completely specified as follows:

Rule Base

1. If x is A^1, then f is b^1.
2. If x is A^2, then f is b^2.
3. If x is A^3, then f is b^3.
4. If x is A^4, then f is b^4.

where $[b^1 \quad b^2 \quad b^3 \quad b^4]^* = [0.5314 \quad 9.0074 \quad 9.5993 \quad 12.6176]$.

Input FuzzySets

$$\mu^1(x) = \exp\left[-\frac{1}{2}\left(\frac{x-1}{1}\right)^2\right]$$

$$\mu^2(x) = \exp\left[-\frac{1}{2}\left(\frac{x-2.5}{1}\right)^2\right]$$

$$\mu^3(x) = \exp\left[-\frac{1}{2}\left(\frac{x-4}{1}\right)^2\right]$$

$$\mu^4(x) = \exp\left[-\frac{1}{2}\left(\frac{x-5.5}{1}\right)^2\right]$$

The premise values of the rules are the same as the input membership values because there is only one input. Therefore, $\mu_i(x) = \mu^i(x)$. The input fuzzy sets are shown in Fig. 8.14.

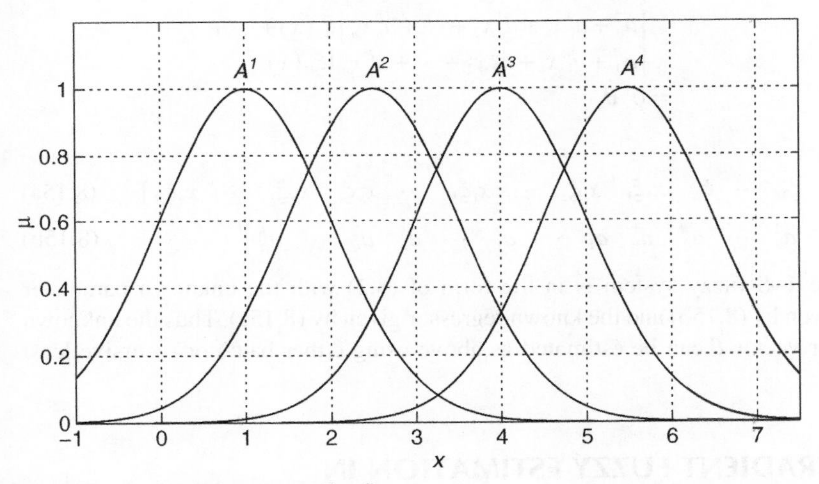

Figure 8.14. Input fuzzy sets (fixed).

Crisp Output

$$y(x) = f(x|\theta) = \frac{\sum_{i=1}^{4} b^i \mu_i(x)}{\sum_{i=1}^{4} \mu_i(x)}$$

In Figure 8.13, the fuzzy characteristic f is the one that results from using $\theta(200)$ in the RLS algorithm.

8.4 LEAST-SQUARES FUZZY ESTIMATION IN TAKAGI–SUGENO FORM

Assume a T–S fuzzy system with R rules in the form

$$\begin{aligned} R_i\text{:} \quad & \text{If } x_1 \text{ is } A_1^K \text{ and } x_2 \text{ is } A_2^L \text{ and } \cdots x_n \text{ is } A_n^M \text{, then} \\ & q^i(x) = a_0^i + a_1^i x_1 + a_2^i x_2 + \cdots + a_n^i x_n \end{aligned} \tag{8.14}$$

The consequent $q^i(x)$ is an affine function of the inputs $x_1, \ldots, x_n$ with the parameters $\{a_0^i, a_1^i, a_2^i, \ldots, a_n^i\}, i = 1, 2, \ldots, R$, unknown. If the premise membership function for Rule i is $\mu_i(x)$, the output of the fuzzy system is

$$
\begin{aligned}
f(x) &= \frac{\sum_{i=1}^{R} q^i \mu_i(x)}{\sum_{i=1}^{R} \mu_i(x)} \\
&= q^1 \xi_1(x) + q^2 \xi_2(x) + \cdots + q^R \xi_R(x) \\
&= \left[a_0^1 + a_1^1 x_1 + a_2^1 x_2 + \cdots + a_n^1 x_n\right] \xi_1(x) + \\
&\quad \left[a_0^2 + a_1^2 x_1 + a_2^2 x_2 + \cdots + a_n^2 x_n\right] \xi_2(x) + \cdots + \\
&\quad \left[a_0^R + a_1^R x_1 + a_2^R x_2 + \cdots + a_n^R x_n\right] \xi_R(x) \\
&= \phi^T \theta
\end{aligned}
$$

where

$$
\phi^T = [\xi_1 \quad \xi_2 \quad \cdots \quad \xi_R \quad x_1\xi_1 \quad x_1\xi_2 \quad \cdots \quad x_1\xi_R \quad \cdots \quad x_n\xi_1 \quad x_n\xi_2 \quad \cdots \quad x_n\xi_R] \tag{8.15a}
$$

$$
\theta = \left[a_0^1 \quad a_0^2 \quad \cdots \quad a_0^R \quad a_1^1 \quad a_1^2 \quad \cdots \quad a_1^R \quad \cdots \quad a_n^1 \quad a_n^2 \quad \cdots \quad a_n^R\right]^T \tag{8.15b}
$$

The T–S fuzzy system is in the form of (8.2) with the unknown parameter vector given by (8.15b) and the known regressor given by (8.15a). Thus the unknown parameter vector θ can be estimated as above using either batch or recursive least squares.

8.5 GRADIENT FUZZY ESTIMATION IN MAMDANI FORM

It may be desirable to adjust input membership function centers and spreads as well as output centers. This may be the case, for example, if we are unsure of what the input fuzzy sets should be. If input memberships are to be adjusted, least-squares methods cannot be used because it is no longer possible to express the system as linear in the parameters. The gradient method [33,34] can be used to adapt both input and output fuzzy sets (although other methods may be used as well [35]).

The basic idea of the gradient method is to move the parameter estimate in a direction that is opposite that of the gradient (or slope) of the error surface in parameter space. This insures the error always decreases with each new parameter update. One problem with the gradient method is that the error surface is not convex so it has many local minima. This means that the converged parameter vector depends on the initial parameter guesses and may not be the one that produces the absolute minimum error. Therefore, if the resulting fuzzy system does not approximate the nonlinear function to sufficient accuracy, it may be beneficial to try different initial parameter guesses.

Assume an n-input Mamdani fuzzy system with R rules, nR input fuzzy sets characterized by Gaussian membership functions with adjustable centers c_j^i and spreads σ_j^i on each input universe (ith rule, jth universe of discourse), and R output singleton fuzzy sets with adjustable locations b^i. Let the rule base of the system be given by:

$$\begin{aligned}
&1.\quad \text{If } x_1 \text{ is } A_1^1 \text{ and } x_2 \text{ is } A_2^1 \text{ and } \cdots \text{ and } x_n \text{ is } A_n^1, \text{ then } y \text{ is } B^1.\\
&2.\quad \text{If } x_1 \text{ is } A_1^2 \text{ and } x_2 \text{ is } A_2^2 \text{ and } \cdots \text{ and } x_n \text{ is } A_n^2, \text{ then } y \text{ is } B^2.\\
&\vdots\\
&R.\quad \text{If } x_1 \text{ is } A_1^R \text{ and } x_2 \text{ is } A_2^R \text{ and } \cdots \text{ and } x_n \text{ is } A_n^R, \text{ then } y \text{ is } B^R.
\end{aligned} \tag{8.16}$$

Notice that these rules are in a slightly different form than we have been considering in that each rule has its own unique premise and consequent fuzzy sets. For instance, the fuzzy sets A_j^1, $j = 1, 2, \ldots, n$ and B^1 in (8.16) are used only for the premise and consequent of Rule 1 and not for any other rule (see Fig. 8.15). This is in contrast to rule bases seen previously in this book, in which a relatively small number of fuzzy sets are defined on each input universe and a rule is written for every possible combination of these.

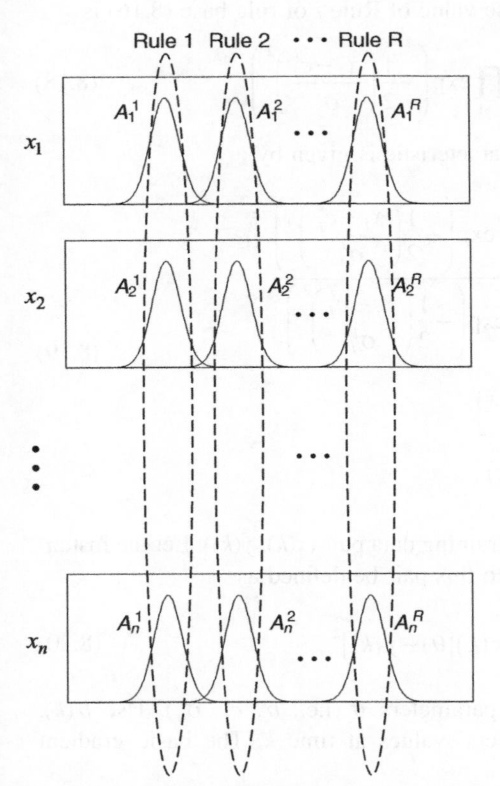

Figure 8.15. Arrangement of inputs and rules corresponding to rule base (8.16). The fuzzy sets are not necessarily in the order shown, that is, all of Rule 1's fuzzy sets are not necessarily leftmost.

Note also that the above form (8.16) for the rule base does not exclude rule bases with rules in the form of (8.8) (e.g., the rule base of the fuzzy controller for the ball and beam of Section 4.1.2). Rule bases with rules in the form of (8.8) tend to be more intuitive for heuristic controller design than the form of (8.16). The form seen in (8.8) can be used when the input fuzzy sets are fixed *a priori* and are not adjusted in the estimation process, such as least-squares estimation methods. On the other hand, rule bases of the form (8.16) must be used in methods where the input memberships are adjusted, such as gradient-based algorithms.

For the rule base of (8.16), assuming we use Gaussian input membership functions, the membership function characterizing A_j^i (ith rule, jth universe) is given by

$$\mu_j^i(x_j) = \exp\left(-\frac{1}{2}\left(\frac{x_j - c_j^i}{\sigma_j^i}\right)^2\right) \tag{8.17}$$

Note that this is a different notation from what we have previously used. In the previous notation, the superscript denoted a particular fuzzy set on a universe [see (3.2)]. In this new notation, the superscript denotes the rule number.

Using *product* T-norm, the premise value of Rule i of rule base (8.16) is

$$\mu_i(x_1, x_2, \cdots, x_n) = \prod_{j=1}^{n} \exp\left(-\frac{1}{2}\left(\frac{x_j - c_j^i}{\sigma_j^i}\right)^2\right) \tag{8.18}$$

Then, the fuzzy system input–output characteristic is given by

$$\begin{aligned} f(x|\theta) &= \frac{\sum_{i=1}^{R} b^i \prod_{j=1}^{n} \exp\left(-\frac{1}{2}\left(\frac{x_j - c_j^i}{\sigma_j^i}\right)^2\right)}{\sum_{i=1}^{R} \prod_{j=1}^{n} \exp\left(-\frac{1}{2}\left(\frac{x_j - c_j^i}{\sigma_j^i}\right)^2\right)} \\ &= \frac{\sum_{i=1}^{R} b^i \mu_i(x)}{\sum_{i=1}^{R} \mu_i(x)} \end{aligned} \tag{8.19}$$

Now assume we are given the kth training data pair $(x(k), y(k))$. Let the instantaneous estimation error corresponding to this pair be defined as

$$e(k) = \frac{1}{2}[f(x(k)|\theta) - y(k)]^2 \tag{8.20}$$

We seek to reduce $e(k)$ by choice of parameters θ (i.e., b^i, c_j^i, σ_j^i). Use $b^i(k)$, $c_j^i(k)$, $\sigma_j^i(k)$ to denote these parameters' values at time k. The basic gradient method of parameter update is

$$b^i(k+1) = b^i(k) - \lambda_1 \frac{\partial e(k)}{\partial b^i(k)} \tag{8.21a}$$

$$c_j^i(k+1) = c_j^i(k) - \lambda_2 \frac{\partial e(k)}{\partial c_j^i(k)} \tag{8.21b}$$

$$\sigma_j^i(k+1) = \sigma_j^i(k) - \lambda_3 \frac{\partial e(k)}{\partial \sigma_j^i(k)} \tag{8.21c}$$

Thus, the new value of a parameter equals the current value plus a correction in the negative direction of the gradient of $e(k)$ with respect to the parameter in parameter space. The constants λ_1, λ_2, and λ_3 are user-defined and determine the size of the step taken in the direction of the negative gradient. In this way, it is insured that $e(k)$ decreases (or at worst does not increase) at every time step.

Note that in this book Gaussian membership functions are always used when gradient parameter update is used, because it is much more straightforward to take derivatives of Gaussians with respect to their parameters than it is to take derivatives of triangulars with respect to theirs.

Calculating the gradients $\partial e(k)/\partial b^i(k)$, $\partial e(k)/\partial c_j^i(k)$, and $\partial e(k)/\partial \sigma_j^i(k)$ is done by applying the chain rule for differentiation. The process is similar to calculation of weight updates via backpropagation for neural networks. To find $\partial e(k)/\partial b^i(k)$, we note from (8.20) that b^i is not mentioned explicitly in $e(k)$, but only implicitly through f. Therefore, we cannot directly calculate $\partial e(k)/\partial b^i(k)$, but instead must do it implicitly using the chain rule of differentiation. Suppressing the time increment k for notational simplicity, we have from (8.20):

$$\frac{\partial e}{\partial b^i} = \frac{\partial e}{\partial f}\frac{\partial f}{\partial b^i}$$

The partial $\partial e/\partial f$ can be obtained directly from (8.20) as

$$\frac{\partial e}{\partial f} = f - y := \varepsilon \tag{8.22}$$

The partial $\partial f/\partial b^i$ can be obtained from (8.9), which can be rewritten as

$$f = b^1 \frac{\mu_1}{\sum \mu_j} + b^2 \frac{\mu_2}{\sum \mu_j} + \cdots + b^R \frac{\mu_R}{\sum \mu_j} \tag{8.23}$$

Therefore, we have

$$\frac{\partial f}{\partial b^i} = \frac{\mu_i}{\sum_{j=1}^{R} \mu_j} \tag{8.24}$$

giving the update law for b^i:

$$b^i(k+1) = b^i(k) - \lambda_1 \varepsilon(k) \frac{\mu_i(x(k))}{\sum_{i=1}^{R} \mu_i(x(k))} \tag{8.25}$$

Similarly, for $\partial e(k)/\partial c_j^i$ we note from (8.20) that c_j^i is not mentioned explicitly in $e(k)$ and not explicitly in f (8.23). However, it is an explicit factor in μ_i which is an explicit factor in f, which is an explicit factor in $e(k)$. Therefore, we calculate $\partial e(k)/\partial c_j^i$ using the chain rule as

$$\frac{\partial e}{\partial c_j^i}=\frac{\partial e}{\partial f}\frac{\partial f}{\partial \mu_i}\frac{\partial \mu_i}{\partial c_j^i}$$

The partial $\partial e/\partial f$ is calculated in (8.22). The partial $\partial f/\partial \mu_i$ is calculated from (8.23) as

$$\frac{\partial f}{\partial \mu_i}=\frac{\partial}{\partial \mu_i}\left(\frac{b^1\mu_1+b^2\mu_2+\cdots+b^R\mu_R}{\mu_1+\mu_2+\cdots+\mu_R}\right)=\frac{\left(\sum_{j=1}^{R}\mu_j\right)b^i-\left(\sum_{j=1}^{R}b^j\mu_j\right)(1)}{\left(\sum_{j=1}^{R}\mu_j\right)^2}$$

$$=\frac{b^i-f}{\sum_{j=1}^{R}\mu_j} \tag{8.26}$$

The partial $\partial \mu_i/\partial c_j^i$ is calculated from (8.18), which can be rewritten as

$$\mu_i(x)=e^{-\frac{1}{2}\left(\frac{x_1-c_1^i}{\sigma_1^i}\right)^2}e^{-\frac{1}{2}\left(\frac{x_2-c_2^i}{\sigma_2^i}\right)^2}\cdots e^{-\frac{1}{2}\left(\frac{x_n-c_n^i}{\sigma_n^i}\right)^2} \tag{8.27}$$

Then

$$\frac{\partial \mu_i}{\partial c_j^i}=\left\{\prod_{k=1}^{j-1}e^{-\frac{1}{2}\left(\frac{x_k-c_k^i}{\sigma_k^i}\right)^2}\right\}\left\{\prod_{k=j+1}^{n}e^{-\frac{1}{2}\left(\frac{x_k-c_k^i}{\sigma_k^i}\right)^2}\right\}e^{-\frac{1}{2}\left(\frac{x_j-c_j^i}{\sigma_j^i}\right)^2}\left(-\left(\frac{x_j-c_j^i}{\sigma_j^i}\right)\right)\left(-\frac{1}{\sigma_j^i}\right)$$

$$=\mu_i\frac{x_j-c_j^i}{(\sigma_j^i)^2} \tag{8.28}$$

Combining these we obtain the update law for c_j^i:

$$c_j^i(k+1)=c_j^i(k)-\lambda_2\varepsilon(k)\left[\frac{b^i(k)-f(x(k)|\theta(k))}{\sum_{i=1}^{R}\mu_i(x(k))}\right]\mu_i(x(k))\left(\frac{x_j(k)-c_j^i(k)}{(\sigma_j^i(k))^2}\right) \tag{8.29}$$

For $\partial e(k)/\partial \sigma_j^i$, we note from (8.20) that σ_j^i is not mentioned explicitly in $e(k)$ and not explicitly in f. However, it is an explicit factor in μ_i which is an explicit factor in f, which is an explicit factor in $e(k)$. Therefore, we calculate $\partial e(k)/\partial \sigma_j^i$ using the chain rule as

$$\frac{\partial e}{\partial \sigma_j^i}=\frac{\partial e}{\partial f}\frac{\partial f}{\partial \mu_i}\frac{\partial \mu_i}{\partial \sigma_j^i}$$

The partials $\partial e/\partial f$ and $\partial f/\partial \mu_i$ are the same as above (8.22) and (8.26). The partial $\partial \mu_i/\partial \sigma_j^i$ is calculated similarly to $\partial \mu_i/\partial c_j^i$ from (8.27) as follows:

$$\frac{\partial \mu_i}{\partial \sigma_j^i} = \left\{ \prod_{k=1}^{j-1} e^{-\frac{1}{2}\left(\frac{x_k - c_k^i}{\sigma_k^i}\right)^2} \right\} \left\{ \prod_{k=j+1}^{n} e^{-\frac{1}{2}\left(\frac{x_k - c_k^i}{\sigma_k^i}\right)^2} \right\} e^{-\frac{1}{2}\left(\frac{x_j - c_j^i}{\sigma_j^i}\right)^2} \left(-\left(\frac{x_j - c_j^i}{\sigma_j^i}\right)\right)\left(-\frac{x_j - c_j^i}{(\sigma_j^i)^2}\right)$$

$$= \mu_i \frac{(x_j - c_j^i)^2}{(\sigma_j^i)^3}$$

By combining these, we obtain the update law for σ_j^i as

$$\sigma_j^i(k+1) = \sigma_j^i(k) - \lambda_3 \varepsilon(k) \left(\frac{b^i(k) - f(x(k)|\theta(k))}{\sum_{i=1}^{R} \mu_i(x(k))} \right) \mu_i(x(k)) \frac{(x_j(k) - c_j^i(k))^2}{(\sigma_j^i(k))^3} \quad (8.30)$$

Equations (8.25), (8.29) and (8.30) give the update laws for the parameters b^i, c_j^i, and σ_j^i of the fuzzy system.

In these update laws, the magnitude of the correction is proportional to the constant λ_i, which is user-selected. If λ_i is small, the rate of convergence is slow, and if λ_i is large, the rate of convergence is fast. Note, however, that if λ_i is too small, convergence may take too long. If λ_i is too large, convergence may not occur at all because each update step in the vicinity of the local minimum is too large to find the minimum.

The gradients in the update laws depend on the training data pairs $(x(k), y(k))$ taken from the system. If several pairs of training data are available, which is usually the case, the above algorithm could be run over and over for the first pair of data until convergence occurs, then iterate again using the second pair of data with the converged parameters of the first set as initial conditions, and so on. Alternatively, we could run through the entire set of training data once for each pair, then iterate over the entire set repeatedly until convergence is obtained. The latter method is used in this text for test data consisting of a finite number of pairs. In adaptive identifiers and controllers, the new data is fed to the algorithm as it occurs.

EXAMPLE 8.4

Consider again the function (8.11). Let 200 data pairs be taken, that is, $\{(x(1), y(1)), (x(2), y(2)), \ldots, (x(200), y(200))\}$, where $x(k)$ is a random variable uniformly distributed in [0 6] for each k. Let there be five rules and five Gaussian input fuzzy sets, one for each rule. Therefore, we have 15 parameters to adjust: five input centers, five input spreads, and five output singleton locations. The initial parameter values are chosen randomly within the range of the training data. Since the input ranges between 0 and 6, we choose five values randomly in this range for initial values of c^i, $i = 1, \ldots, 5$. Similarly, we choose five values randomly in [0 15] for initial values of b^i, $i = 1, \ldots, 5$. The initial values of σ^i, $i = 1, \ldots, 5$, are randomly chosen from [0 2]. Note that if we have some idea of the correct parameter values, we should use these as initial conditions for the gradient algorithm. The step sizes are chosen as $\lambda_1 = \lambda_2 = \lambda_3 = 0.001$.

The characteristic of the converged fuzzy system after 400 training steps (i.e., twice through the training data) is shown in Figure 8.16. The fit is not good after so few training steps. The evolution of the parameter estimates is shown in Figures 8.17–8.19. From these figures, it is evident that convergence has not occurred after 400 training steps.

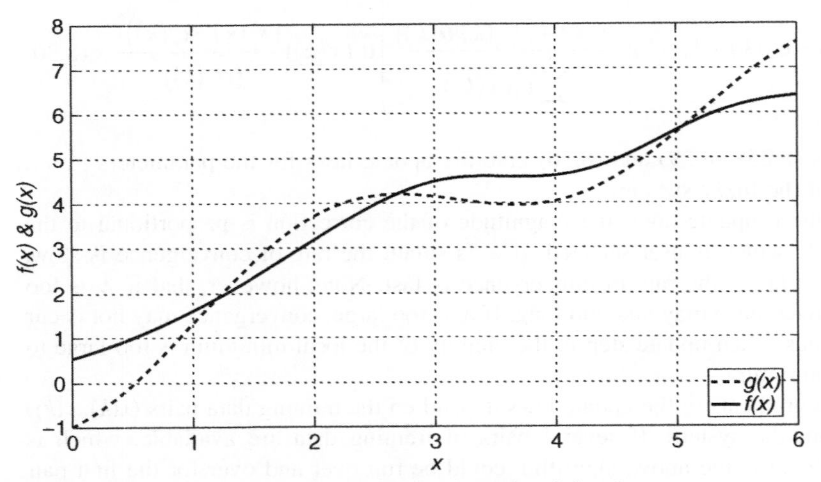

Figure 8.16. Comparison of f and g, twice through training data.

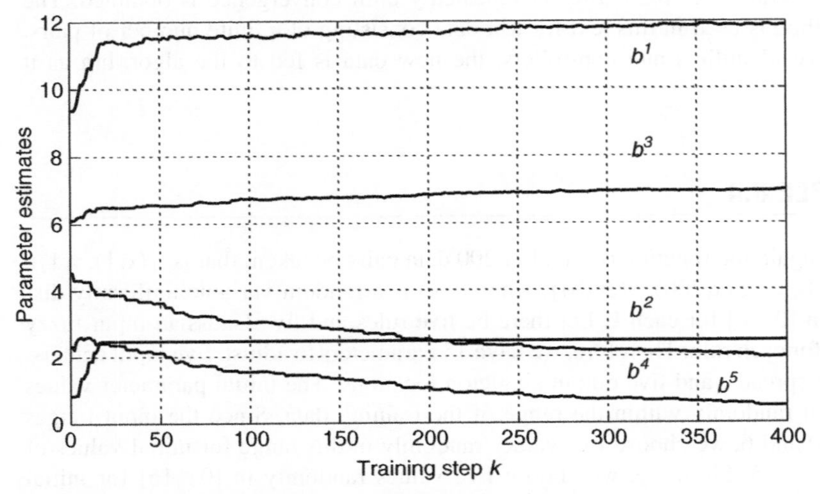

Figure 8.17. Estimates of output membership function centers.

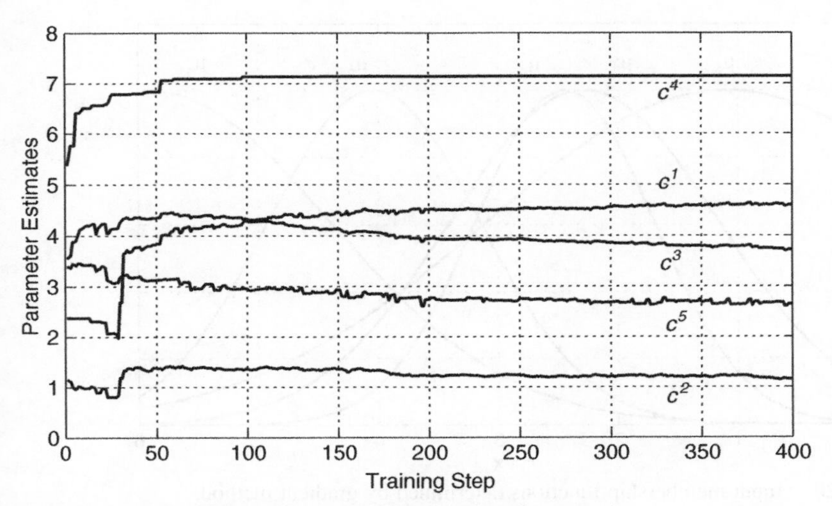

Figure 8.18. Estimates of input membership function centers.

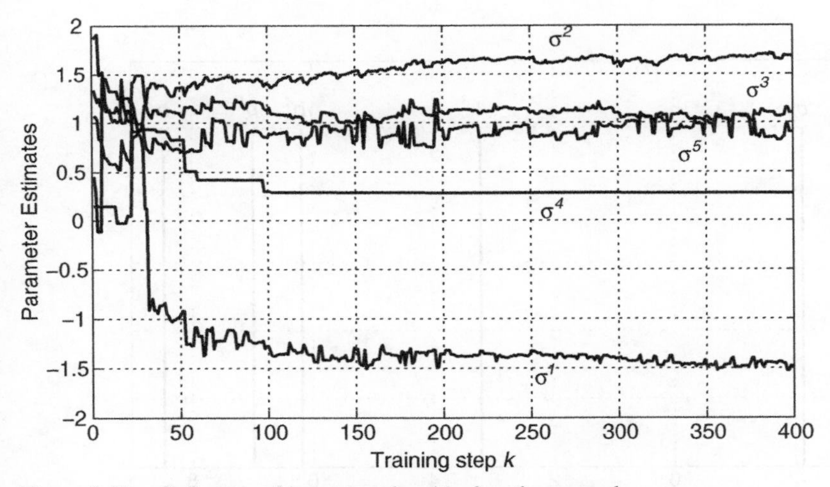

Figure 8.19. Estimates of input membership function spreads.

If we perform 10^5 training steps (i.e., 500 times through the training data), the estimates have converged and the fuzzy characteristic $f(x)$ is indistinguishable from $g(x)$, that is, the approximation is excellent.

After 10^5 training steps, the input centers and spreads have converged to $c^* = [5.8952 \quad 2.4475 \quad 4.0362 \quad 2.2839 \quad 1.0931]$ and $\sigma^* = [1.4974 \quad 0.7503 \quad 0.8877 \quad 1.1236 \quad 2.0125]$, producing the input memberships shown in Figure 8.20.

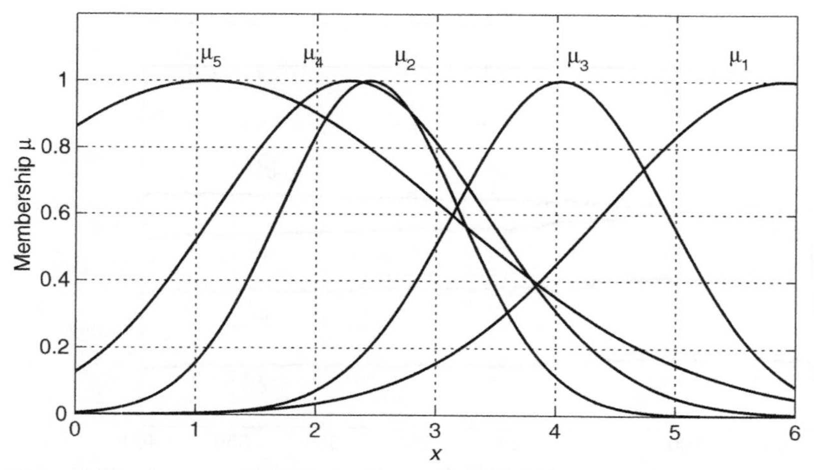

Figure 8.20. Input membership functions determined by gradient method.

The output singleton locations have converged to $b^* = [8.4704 \quad 6.9878 \quad 3.2109 \quad 6.1789 \quad -2.1289]$, producing the output memberships shown in Figure 8.21.

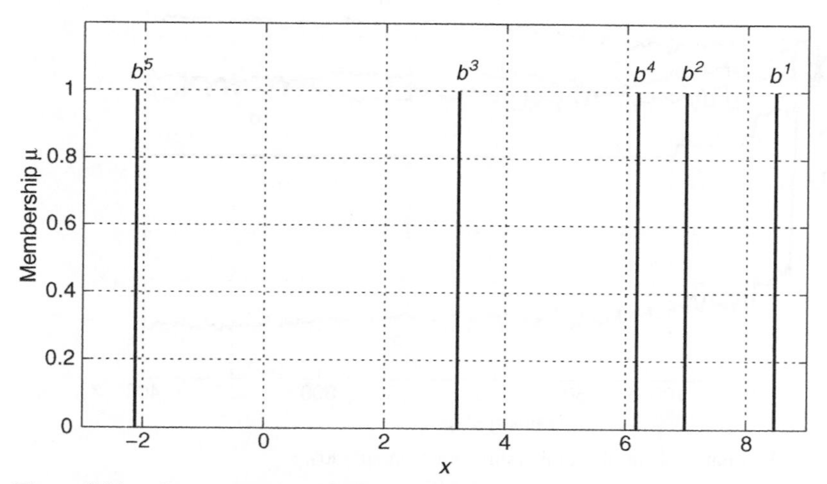

Figure 8.21. Output membership functions determined by gradient method.

Thus, the rule base of the fuzzy system approximating $g(x)$ in (8.11) is

1. If x is A^1, then f is b^1.
2. If x is A^2, then f is b^2.
3. If x is A^3, then f is b^3.
4. If x is A^4, then f is b^4.
5. If x is A^5, then f is b^5.

Note that for each different set of initial conditions and training data, the gradient algorithm produces a different fuzzy system, that is, different input and output memberships. This is due to the fact that the error function for the gradient method does not have one unique minimum, but rather many local minima. For most reasonable initial parameter guesses and training data, the gradient algorithm will eventually converge to a fuzzy system whose input–output characteristic closely matches $g(x)$. Each fuzzy system is different, depending on the initial parameter guesses and training data used. A good fit to $g(x)$ is not guaranteed, however.

8.6 GRADIENT FUZZY ESTIMATION IN TAKAGI–SUGENO FORM

Assume an n-input T–S fuzzy system with R rules, nR input fuzzy sets characterized by Gaussian membership functions with adjustable centers c_j^i and spreads σ_j^i on each input universe (ith rule, jth universe of discourse), and R consequents consisting of affine functions with n + 1 adjustable parameters a_j^i. Let the rule base be given by

$$\begin{aligned}
&1.\ \text{If } x_1 \text{ is } A_1^1 \text{ and } x_2 \text{ is } A_2^1 \text{ and } \cdots \text{ and } x_n \text{ is } A_n^1 \text{, then} \\
&\quad q^1(x) = a_0^1 + a_1^1 x_1 + a_2^1 x_2 + \cdots a_n^1 x_n \\
&2.\ \text{If } x_1 \text{ is } A_1^2 \text{ and } x_2 \text{ is } A_2^2 \text{ and } \cdots \text{ and } x_n \text{ is } A_n^2 \text{, then} \\
&\quad q^2(x) = a_0^2 + a_1^2 x_1 + a_2^2 x_2 + \cdots a_n^2 x_n \\
&\quad \vdots \\
&R.\ \text{If } x_1 \text{ is } A_1^R \text{ and } x_2 \text{ is } A_2^R \text{ and } \cdots \text{ and } x_n \text{ is } A_n^R \text{, then} \\
&\quad q^R(x) = a_0^R + a_1^R x_1 + a_2^R x_2 + \cdots a_n^R x_n
\end{aligned} \tag{8.31}$$

The crisp output of this system is given by

$$f(x) = \frac{\sum_{j=1}^{R} q^i(x)\mu_i(x)}{\sum_{j=1}^{R} \mu_i(x)}$$

where $\mu_i(x)$ are the premise membership functions defined in (8.18).

Now the parameters of the system are c_j^i, σ_j^i, and a_j^i. These are to be adjusted using the gradient method. The adjustment of c_j^i and σ_j^i are the same and before [(8.29) and (8.30)]. The update law for a_j^i is

$$a_j^i(k+1) = a_j^i(k) - \lambda_4 \frac{\partial e(k)}{\partial a_j^i(k)} \tag{8.32}$$

The partial $\partial e(k)/\partial a_j^i$ is calculated using the chain rule for differentiation as above. We note from (8.20) that a_j^i is not mentioned explicitly in $e(k)$, but only implicitly through f. Therefore, we cannot directly calculate $\partial e(k)/\partial a_j^i$, but instead must do it implicitly using the chain rule of differentiation:

$$\frac{\partial e}{\partial a_j^i} = \frac{\partial e}{\partial f} \frac{\partial f}{\partial q^i} \frac{\partial q^i}{\partial a_j^i}$$

The partial $\partial e/\partial f$ was calculated in (8.22). The partial $\partial f/\partial q^i$ can be obtained from (8.23), which can be rewritten as

$$f = q^1 \frac{\mu_1}{\sum \mu_j} + q^2 \frac{\mu_2}{\sum \mu_j} + \cdots + q^R \frac{\mu_R}{\sum \mu_j}$$

This yields

$$\frac{\partial f}{\partial q^i} = \frac{\mu_i}{\sum \mu_i}$$

The partial $\partial q^i / \partial a_j^i$ depends on j, that is,

$$\frac{\partial q^i(x)}{\partial a_j^i} = \begin{cases} 1, & j = 0 \\ x_j, & j = 1, \cdots, n \end{cases} \tag{8.33}$$

Therefore, the parameter update law for a_j^i is

$$a_j^i(k+1) = a_j^i(k) - \lambda_4 \varepsilon(k) \frac{\mu_i(x(k))}{\sum_{j=1}^{R} \mu_i(x(k))} \frac{\partial q^i(x(k),k)}{\partial a_j^i(k)} \tag{8.34}$$

with $\partial q^i / \partial a_j^i$ as in (8.33). Then, the parameter update laws for T–S fuzzy systems with affine consequents are given by (8.29), (8.30), and (8.34).

8.7 SUMMARY

This chapter explores the ability of fuzzy systems to estimate static functions. To do this, input–output data must be available from the function. Although there are many methods that can be used to create fuzzy systems that aproximate functions, we concentrate on two: least squares and gradient. This is because these methods are recursive, hence can be used for adaptive fuzzy control (Chapter 10).

In the case of **least-squares** estimation (whether batch or recursive), the fuzzy system must be linear in the unknown parameters. Therefore, the input membership functions must be fixed *a priori* and only the output parameters adjusted. Because the error function is quadratic, it has a unique minimum in parameter space. Therefore, the parameter estimates converge to unique values regardless of the initial guess.

The recursive least-squares algorithm converges more quickly than some other methods, specifically the gradient algorithm. Because least squares is limited to adjusting only the output fuzzy sets, it is less flexible in constructing fuzzy systems to estimate nonlinear functions than methods that adjust both input and output fuzzy sets, specifically the gradient algorithm.

If least squares is used, it is possible to use rules of the form found in Section 3.6 (wind chill system). This enables one to define intuitively a relatively small number of fuzzy sets on each input universe and write rules whose premises are various combinations of these. This produces a rule base that is transparent to human understanding.

In the case of **gradient** adaptation, there is no requirement for the fuzzy system to be linear in the parameters, so all input and output fuzzy sets can adjusted. Because of this, each rule must have its own separate input fuzzy sets that are not shared with any other rules. This necessitates a premise fuzzy set arrangement as in (8.16) rather than that found in Section 3.6. Thus gradient-adapted rules are not as easily understood as least-squares-adapted rules.

Gradient estimation takes longer to converge and may not converge at all in a finite time interval. However, gradient estimation tends to be more flexible in its estimation capabilities, due to the fact that the input memberships are adapted along with the output parameters.

EXERCISES

8.1 In a generalization of the least squares algorithm presented in Section 8.1.1, some data is weighted heavier than other data giving rise to the performance measure

$$V_2(\hat{\theta}) = \frac{1}{2}\sum_{j=1}^{k} w_j\left[y(j) - \phi^T(j)\hat{\theta}\right]^2$$

where $w_j > 0$, $j = 1, \ldots, k$. Derive the value of $\hat{\theta}^*$ that minimizes V_2 (this is called a *weighted least-squares* estimate).

8.2 Repeat Example 8.1 with 5 input fuzzy sets centered at [0 1.5 3 4.5 6]. Choose σ so the adjacent input membership functions cross at 0.5.

8.3 Derive the recursive least-squares algorithm (8.7) from the batch least-squares estimate (8.6) by writing $\hat{\theta}^*(k)$ in terms of $\hat{\theta}^*(k-1)$, defining $P(k) = [\Phi^T(k)\Phi(k)]^{-1}$, and using the Matrix Inversion Lemma.

8.4 Repeat Example 8.2 using data taken at points $(x_1, x_2) = \{(-1\ \ -1), (-1, 0), (-1, 1), (0, -1), (0, 0), (0, 1), (1, -1), (1, 0), (1, 1)\}$

8.5 Use RLS to find a two-rule T–S fuzzy system to approximate (8.11) on the intertval $0 \le x \le 6$. Let the rules be

1. If x is A^1, then $q^1(x) = a_0^1 + a_1^1 x$.
2. If x is A^2, then $q^2(x) = a_0^2 + a_1^2 x$.

Use triangular membership functions forming a partition of unity on the x universe.

8.6 You are using RLS to find a four-rule T–S fuzzy system to approximate (8.13) on the intertval $-1 \le x_1 \le 1$, $-1 \le x_2 \le 1$, as in Section 8.4. (a) Specify all rules of the approximating fuzzy system. (b) Specify all fuzzy sets. (c) Specify the T-norm and method of defuzzification. (d) Specify the training data and test data.

8.7 (a) Why are rules of the form found in Sect. 3.6 more desirable than rules of the form (8.16)? (b) Why can rules like those in Sect. 3.6 not be used in gadient-adjusted fuzzy systems if input membership functions are to be adjusted?

8.8 Redo Example 8.3 with 3 Gaussian input fuzzy sets with centers at $x = [0, 3, 6]$ such that adjacent memberships cross at 0.5.

8.9 Redo problem 8.8 with 3 triangular memberships located at $x = [0, 3, 6]$ and forming a partition of unity.

8.10 Assume a fuzzy system with rules of the form

$$R_i: \quad \text{If } x \text{ is } A^i\text{, then } q^i(x) = a_0^i \sin(x) + a_1^i \cos(x)$$

is to be adjusted via gradient. Write the gradient update equations for a_0^i and a_1^i.

CHAPTER 9

MODELING OF DYNAMIC PLANTS AS FUZZY SYSTEMS

The estimation methods of Chapter 8 can be used to find fuzzy models for nonlinear dynamic systems. The determination of a model for an unknown system is known as *identification*. Usually, the object of doing so is to utilize the model in a parallel distributed control scheme (see Chapter 7). Design of a parallel distributed controller is relatively straightforward because the consequents in the fuzzy model are linear systems.

In Section 9.1, we present a method of deriving a fuzzy model for a nonlinear system whose equations of motion are known. This enables parallel distributed control of the nonlinear system when derivation of the controller directly from the nonlinear equations of motion may he impossible.

If the controller is to be capable of adjusting itself in real time (as in adaptive fuzzy control), the method used for identification must be recursive. In Chapter 8, we give two methods of recursive identification: recursive least squares (RLS) and gradient. Therefore, we will concentrate on these methods in this chapter.

9.1 MODELING KNOWN PLANTS AS T–S FUZZY SYSTEMS [29]

Many nonlinear systems with known mathematical models may be exactly modeled on a bounded domain in the state space with Takagi–Sugeno (T–S) fuzzy systems. Consider a nonlinear time-invariant system for which $x = 0$ is the equilibrium point. The basic idea of modeling it with a T–S fuzzy system is to express it as a series of linear dynamic systems, each one the consequent of one of the rules. Functions z_i are determined such that nonlinear terms in the plant model can be expressed as $z_i x$, where x is a plant state. Each of these functions z_i becomes an input to the fuzzy system. Two fuzzy sets forming a partition of unity are defined on each z_i universe, centered at the maximum and minimum of z_i.

Assume a scalar nonlinear function $z(x)$. If $z(x)$ is to be modeled in the domain $\mathcal{X}$, let $b_m = \min_{\mathcal{X}}(z)$ and $b_M = \max_{\mathcal{X}}(z)$. Then, create two fuzzy sets P^1 and P^2 on $\mathcal{X}$ characterized by triangular membership functions $\mu^1(z)$ and $\mu^2(z)$ as

$$\mu^1(z) = \frac{b_M - z}{b_M - b_m} \tag{9.1a}$$

Fuzzy Control and Identification, By John H. Lilly

$$\mu^2(z) = \frac{z - b_m}{b_M - b_m} \tag{9.1b}$$

Then, z can be exactly represented on $\mathcal{X}$ as

$$z = \mu^1(z)b_m + \mu^2(z)b_M \tag{9.2}$$

This technique can be used to exactly model a nonlinear dynamic system as a weighted average of linear systems.

EXAMPLE 9.1

Consider the nonlinear system $\dot{x} = f(x)$ and assume that $f(x)$ can be factored as $z(x)x$ [e.g., x^3 can be factored as $(x^2)x$ so that $z(x) = x^2$]. Now the nonlinear system can be expressed as $\dot{x} = z(x)x$. In order to exactly model the system on the bounded domain $\mathcal{X}$ as a T–S fuzzy system, let $b_m = \min_{\mathcal{X}}(z)$ and $b_M = \max_{\mathcal{X}}(z)$.

Create two fuzzy sets MIN and MAX on $\mathcal{X}$ characterized by triangular membership functions $\mu^{\text{MIN}}(z)$ and $\mu^{\text{MAX}}(z)$, as in (9.1). These are shown in Figure 9.1. Then the nonlinear system is exactly modeled on $\mathcal{X}$ by the T–S fuzzy system:

1. If z is MIN, then $\dot{x}^1 = b_m x$.
2. If z is MAX, then $\dot{x}^2 = b_M x$.

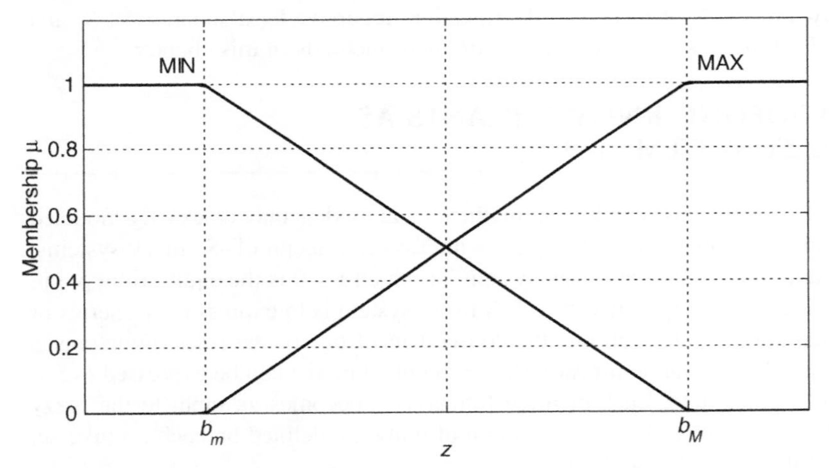

Figure 9.1. Fuzzy sets on z universe.

EXAMPLE 9.2

A nonlinear system has the mathematical model:

$$\dot{x}_1 = x_1 x_2 \tag{9.3a}$$

$$\dot{x}_2 = x_1 - x_2^3 + \left(1 + \cos^2 x_1\right)u \tag{9.3b}$$

It can be verified that this system is open-loop unstable [21,22]. Define $z_1 = x_1$, $z_2 = x_2^2$, and $z_3 = 1 + \cos^2 x_1$. Then, (9.3) can be rewritten as

$$\begin{bmatrix} \dot{x}_1 \\ \dot{x}_2 \end{bmatrix} = \begin{bmatrix} 0 & z_1 \\ 1 & -z_2 \end{bmatrix} \begin{bmatrix} x_1 \\ x_2 \end{bmatrix} + \begin{bmatrix} 0 \\ z_3 \end{bmatrix} u$$

We wish to derive a T–S fuzzy system whose behavior exactly duplicates (9.3). Assume the bounded domain $\mathcal{X}$ is defined by $x_1 \in [-2, 2]$ and $x_2 \in [-2, 2]$ [i.e., the T–S fuzzy system will duplicate (9.3) in this domain]. Then, $\min_{\mathcal{X}} z_1 = b_{1m} = -2$, $\max_{\mathcal{X}} z_1 = b_{1M} = 2$, $\min_{\mathcal{X}} z_2 = b_{2m} = 0$, $\max_{\mathcal{X}} z_2 = b_{2M} = 4$, $\min_{\mathcal{X}} z_3 = b_{3m} = 1$, $\max_{\mathcal{X}} z_3 = b_{3M} = 2$.

For z_1, z_2, and z_3, (9.1) yields

$$\mu^{\text{NEG}}(z_1) = \frac{2 - z_1}{4}$$
$$\mu^{\text{POS}}(z_1) = \frac{z_1 + 2}{4}$$
$$\mu^{\text{SMALL}}(z_2) = \frac{4 - z_2}{4}$$
$$\mu^{\text{LARGE}}(z_2) = \frac{z_2}{4}$$
$$\mu^{\text{SMALL}}(z_3) = \frac{2 - z_3}{1}$$
$$\mu^{\text{LARGE}}(z_3) = \frac{z_3 - 1}{1}$$

These memberships characterize fuzzy sets on the z_1 (Fig. 9.2), z_2 (Fig. 9.3), and z_3 (Fig. 9.4), universes.

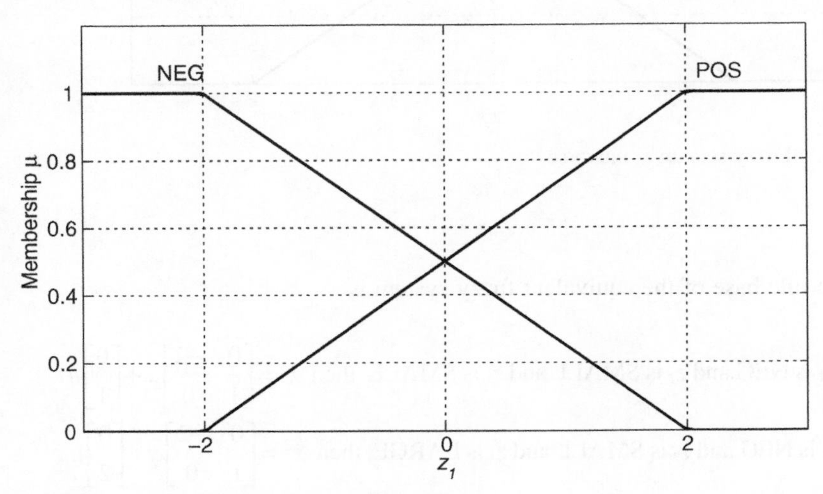

Figure 9.2. Fuzzy sets on z_1 universe.

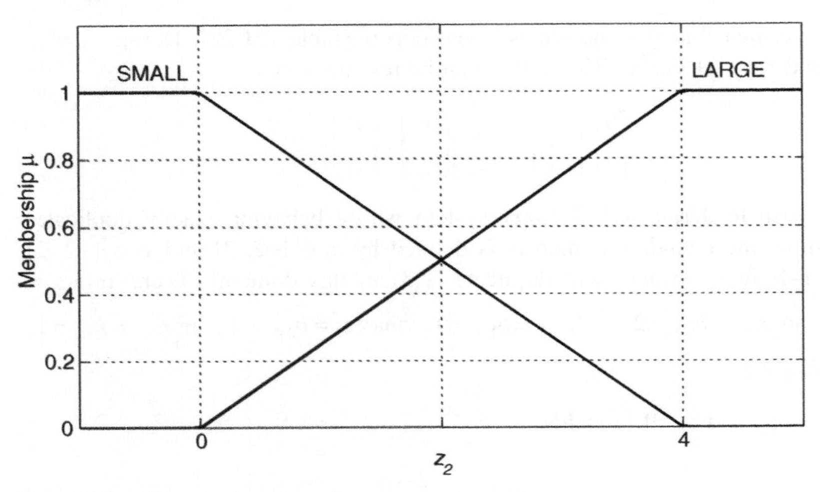

Figure 9.3. Fuzzy sets on z_2 universe.

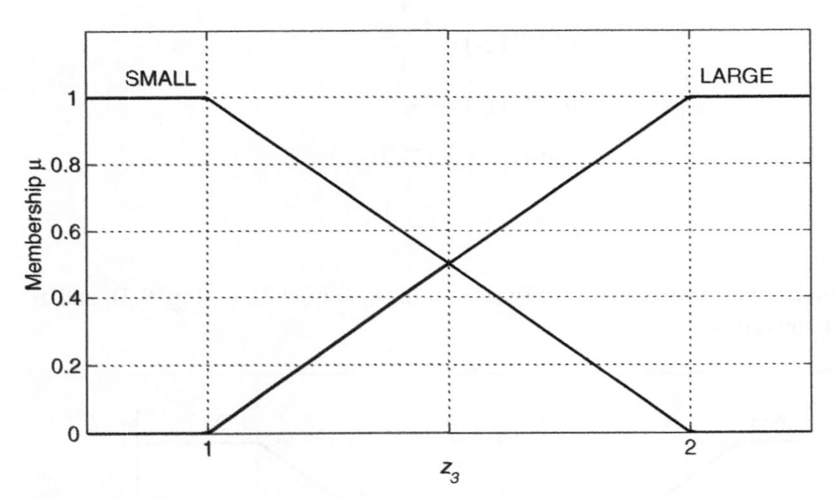

Figure 9.4. Fuzzy sets on z_3 universe.

The rule base of the equivalent fuzzy system is

1. If z_1 is NEG and z_2 is SMALL and z_3 is SMALL, then $\dot{x}^1 = \begin{bmatrix} 0 & -2 \\ 1 & 0 \end{bmatrix} x + \begin{bmatrix} 0 \\ 1 \end{bmatrix} u.$

2. If z_1 is NEG and z_2 is SMALL and z_3 is LARGE, then $\dot{x}^2 = \begin{bmatrix} 0 & -2 \\ 1 & 0 \end{bmatrix} x + \begin{bmatrix} 0 \\ 2 \end{bmatrix} u.$

3. If z_1 is NEG and z_2 is LARGE and z_3 is SMALL, then $\dot{x}^3 = \begin{bmatrix} 0 & -2 \\ 1 & -4 \end{bmatrix} x + \begin{bmatrix} 0 \\ 1 \end{bmatrix} u$.

4. If z_1 is NEG and z_2 is LARGE and z_3 is LARGE, then $\dot{x}^4 = \begin{bmatrix} 0 & -2 \\ 1 & -4 \end{bmatrix} x + \begin{bmatrix} 0 \\ 2 \end{bmatrix} u$.

5. If z_1 is POS and z_2 is SMALL and z_3 is SMALL, then $\dot{x}^5 = \begin{bmatrix} 0 & 2 \\ 1 & 0 \end{bmatrix} x + \begin{bmatrix} 0 \\ 1 \end{bmatrix} u$.

6. If z_1 is POS and z_2 is SMALL and z_3 is LARGE, then $\dot{x}^6 = \begin{bmatrix} 0 & 2 \\ 1 & 0 \end{bmatrix} x + \begin{bmatrix} 0 \\ 2 \end{bmatrix} u$.

7. If z_1 is POS and z_2 is LARGE and z_3 is SMALL, then $\dot{x}^7 = \begin{bmatrix} 0 & 2 \\ 1 & -4 \end{bmatrix} x + \begin{bmatrix} 0 \\ 1 \end{bmatrix} u$.

8. If z_1 is POS and z_2 is LARGE and z_3 is LARGE, then $\dot{x}^8 = \begin{bmatrix} 0 & 2 \\ 1 & -4 \end{bmatrix} x + \begin{bmatrix} 0 \\ 2 \end{bmatrix} u$.

This T–S fuzzy model exactly duplicates (9.3) in $x_1 \in [-2, 2]$, $x_2 \in [-2, 2]$. Note that the $1 + \cos^2 x_1$ term does not have to be expressed as linear in x because it multiplies the input u and not a state.

Now that we have a T–S fuzzy model of (9.3), we can design a parallel distributed controller as in Section 7.1. Let us design a PDC that places the closed-loop eigenvalues of each consequent of the plant fuzzy system at $-5 \pm j5$. Then the required controller fuzzy system has the rule base

1. If z_1 is NEG and z_2 is SMALL and z_3 is SMALL, then $u^1(t) = -[-24 \quad 10]x(t)$.
2. If z_1 is NEG and z_2 is SMALL and z_3 is LARGE, then $u^2(t) = -[-12 \quad 5]x(t)$.
3. If z_1 is NEG and z_2 is LARGE and z_3 is SMALL, then $u^3(t) = -[-24 \quad 6]x(t)$.
4. If z_1 is NEG and z_2 is LARGE and z_3 is LARGE, then $u^4(t) = -[-12 \quad 3]x(t)$.
5. If z_1 is POS and z_2 is SMALL and z_3 is SMALL, then $u^5(t) = -[26 \quad 10]x(t)$.
6. If z_1 is POS and z_2 is SMALL and z_3 is LARGE, then $u^6(t) = -[13 \quad 5]x(t)$.
7. If z_1 is POS and z_2 is LARGE and z_3 is SMALL, then $u^7(t) = -[26 \quad 6]x(t)$.
8. If z_1 is POS and z_2 is LARGE and z_3 is LARGE, then $u^8(t) = -[13 \quad 3]x(t)$.

To prove asymptotic stability, Theorem 7.1 requires that one positive definite symmetric G be bond to satisfy the 64 linear matrix inequalities (7.4) for all $i = 1, \ldots, 8$ and $j = 1, \ldots, 8$. However, it can be verified via simulation (though this is not a proof!) that the above fuzzy PDC controller does render $(x_1, x_2) = (0, 0)$ of (9.3) asymptotically stable.

9.2 IDENTIFICATION IN INPUT–OUTPUT DIFFERENCE EQUATION FORM

Assume an unknown plant with single input u and single output y. In order to find a fuzzy system to model it, it must be excited by an input while data $\{y(k), u(k)\}$, $k = 1, 2, 3, \ldots$ is taken from the plant. If the plant is continuous-time, which is usually the case, samples of the signals are taken every Δt s. In that case, $t = k\Delta t$ with $k = 1, 2, 3, \ldots$. Usually the Δt is suppressed and, for example, $y(k\Delta t)$ is written simply as $y(k)$.

The input sequence $u(k)$ should have sufficient frequency content to identify the plant. In general, the more complex the plant, the more different frequencies should be contained in u. It is generally not known what types of inputs provide sufficient excitation for nonlinear systems, but inputs consisting of many frequencies have the best chance of success.

If the plant is unknown, we can attempt to estimate it as an R-rule T–S fuzzy system with rules in the form:

$$\begin{aligned} R_i: \quad & \text{If } y(k) \text{ is } A_1^K \text{ and } y(k-1) \text{ is } A_2^L \text{ and} \cdots \text{and } y(k-n+1) \text{ is } A_n^M, \text{ then} \\ & y^i(k+1) = a_1^i y(k) + a_2^i y(k-1) + \cdots + a_n^i y(k-n+1) + \\ & \qquad b_1^i u(k) + b_2^i u(k-1) + \cdots + b_n^i u(k-n+1) \end{aligned} \tag{9.4}$$

where the a and b coefficients and, in the case of gradient identification the input fuzzy sets, are to be determined. We note that (9.4) is not the only possibility of model structure in the consequent. The number of past inputs and outputs in the RHS regressions could be different from each other, and the delay of the model, which equals 1 in (9.4), could be other than 1.

9.2.1 Batch Least-Squares Identification in Input–Output Difference Equation Form

If least squares is to be used to identify a fuzzy model to estimate the plant, the input fuzzy sets must be fixed *a priori* so that the fuzzy system can be expressed in a form that is linear in the parameters (see Section 8.1). Therefore, the fuzzy basis functions are also known. If the known premise value of Rule i at time k is $\mu_i(y(k), \ldots, y(k-n+1))$ and defining the known fuzzy basis functions

$$\xi_i(k) = \frac{\mu_i(y(k), \cdots, y(k-n+1))}{\sum_{i=1}^{R} \mu_i(y(k), \cdots, y(k-n+1))} \tag{9.5}$$

for $i = 1, \ldots, R$, the output of the fuzzy system can be expressed as

$$\begin{aligned} \hat{y}(k+1) = & \left[a_1^1 y(k) + \cdots + a_n^1 y(k-n+1) + b_1^1 u(k) + \cdots + b_n^1 u(k-n+1)\right]\xi_1(k) + \\ & \left[a_1^2 y(k) + \cdots + a_n^2 y(k-n+1) + b_1^2 u(k) + \cdots + b_n^2 u(k-n+1)\right]\xi_2(k) + \cdots + \\ & \left[a_1^R y(k) + \cdots + a_n^R y(k-n+1) + b_1^R u(k) + \cdots + b_n^R u(k-n+1)\right]\xi_R(k) \end{aligned} \tag{9.6}$$

or $\hat{y}(k+1) = \phi^{\mathrm{T}}(k)\theta$, where

$$\phi(k)=\begin{bmatrix} y(k)\xi_1(k) \\ \vdots \\ y(k)\xi_R(k) \\ \vdots \\ y(k-n+1)\xi_1(k) \\ \vdots \\ y(k-n+1)\xi_R(k) \\ u(k)\xi_1(k) \\ \vdots \\ u(k)\xi_R(k) \\ \vdots \\ u(k-n+1)\xi_1(k) \\ \vdots \\ u(k-n+1)\xi_R(k) \end{bmatrix} \quad \text{and} \quad \theta=\begin{bmatrix} a_1^1 \\ \vdots \\ a_1^R \\ \vdots \\ a_n^1 \\ \vdots \\ a_n^R \\ b_1^1 \\ \vdots \\ b_1^R \\ \vdots \\ b_n^1 \\ \vdots \\ b_n^R \end{bmatrix} \tag{9.7}$$

From M input–output measurements, form matrices

$$\Phi=\begin{bmatrix} \phi^T(n) \\ \phi^T(n+1) \\ \vdots \\ \phi^T(M-1) \end{bmatrix} \sim (M-n)\times 2nR$$

and

$$Y=[y(n+1) \quad y(n+2) \quad \cdots \quad y(M)]^T \sim (M-n)\times 1$$

Then the batch least-squares estimate of θ based on the M measurements taken is calculated as in (8.6).

EXAMPLE 9.3

Consider the motor-driven robotic link described in Section 1.4.3 and modeled by the truth model [19]:

$$\ddot{\psi} = -64\sin\psi - 5\dot{\psi} + 4\times 10^4 u \tag{9.8}$$

where ψ is the angle of the link from the vertical-down position and u is the current input to the motor. The output is the measured link angle $y(t) = \psi(t)$. The system is simulated with a fourth-order Runge–Kutta integration algorithm with a step size of 0.001 s. For the input–output data, we save pairs $\{u(k\Delta t), y(k\Delta t)\}$ for $k = 1, \ldots, 998$ and $\Delta t = 0.01$ s, that is, we save input and output signals every 10 steps through the RK4 integration routine.

The input function chosen for the identification is a Gaussian pseudorandom sequence with zero mean and a standard deviation of 2×10^{-2}, giving the output shown in Figure 9.5.

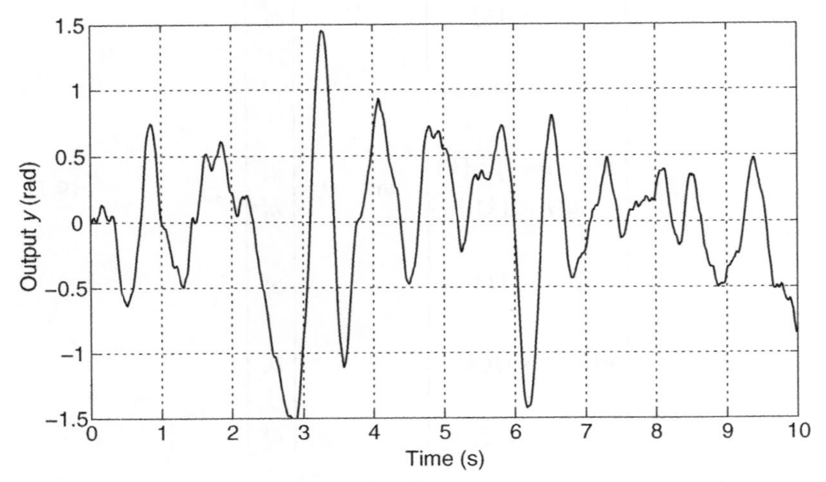

Figure 9.5. Output portion of training data.

Deline m input fuzzy sets characterized by triangular membership functions equally spaced on the universe $-\pi/2 \le y \le \pi/2$ rad and forming partitions of unity. By trying various numbers of inputs n and input fuzzy sets m, it is determined that a sufficient number of inputs to the fuzzy system is $n = 2$, with $m = 2$ fuzzy sets on each input universe. Therefore, the number of rules is $R = 4$, and there are 16 unknown parameters to be determined via batch least squares. This results in $\Phi \sim 996 \times 16$ and $Y \sim 996 \times 1$. The fuzzy sets on each input universe are shown in Figure 9.6.

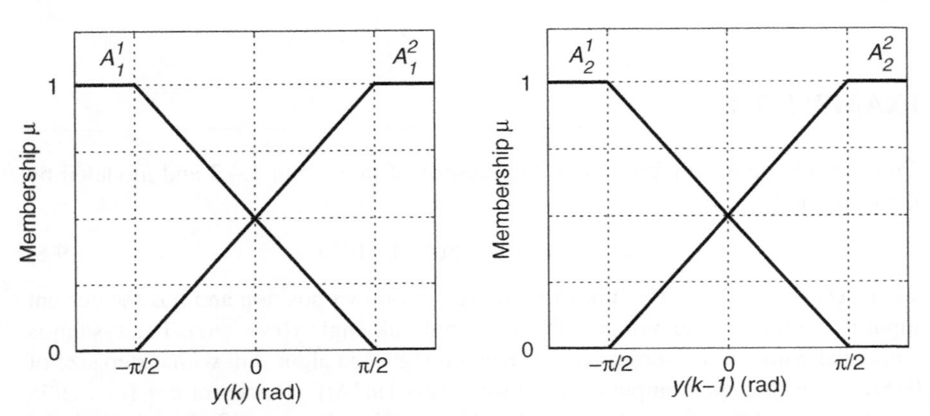

Figure 9.6. Input fuzzy sets for T–S fuzzy system approximating motor-driven robotic link.

The corresponding batch least-squares estimate of consequent parameters from (8.6) is

$$\theta^* = \begin{bmatrix} a_1^{1*} \\ a_1^{2*} \\ a_1^{3*} \\ a_1^{4*} \\ a_2^{1*} \\ a_2^{2*} \\ a_2^{3*} \\ a_2^{4*} \\ b_1^{1*} \\ b_1^{2*} \\ b_1^{3*} \\ b_1^{4*} \\ b_2^{1*} \\ b_2^{2*} \\ b_2^{3*} \\ b_2^{4*} \end{bmatrix} = \begin{bmatrix} 2.46 \\ 1.19 \\ 2.52 \\ 1.43 \\ -1.58 \\ -1.13 \\ -0.735 \\ -0.417 \\ 0.0494 \\ -0.601 \\ 0.929 \\ -0.0703 \\ -0.0314 \\ 1.41 \\ -1.35 \\ -0.0766 \end{bmatrix}$$

Thus the T–S fuzzy system approximating the motor-driven robotic link is given by the rule base:

1. If $y(k)$ is A_1^1 and $y(k-1)$ is A_2^1, then

$$\hat{y}^1(k+1) = 2.46y(k) - 1.58y(k-1) + 0.0494u(k) - 0.0314u(k-1)$$

2. If $y(k)$ is A_1^1 and $y(k-1)$ is A_2^2, then

$$\hat{y}^2(k+1) = 1.19y(k) - 1.13y(k-1) - 0.601u(k) + 1.41u(k-1)$$

3. If $y(k)$ is A_1^2 and $y(k-1)$ is A_2^1, then

$$\hat{y}^3(k+1) = 2.52y(k) - 0.735y(k-1) + 0.929u(k) - 1.35u(k-1)$$

4. If $y(k)$ is A_1^2 and $y(k-1)$ is A_2^2, then

$$\hat{y}^4(k+1) = 1.43y(k) - 0.417y(k-1) - 0.0703u(k) - 0.0766u(k-1)$$

where the fuzzy sets A_1^1, A_1^2, A_2^1, and A_2^2 are specified in Figure 9.6. The output of the fuzzy system is

$$\hat{y}(k+1)) = \frac{\sum_{j=1}^{4} \hat{y}^j \mu_j(y(k), y(k-1))}{\sum_{j=1}^{4} \mu_j(y(k), y(k-1))}$$

where the premise membership values μ_j, $j = 1, \ldots, 4$ are calculated as

$$\mu_1(y(k), y(k-1)) = \mu_1^1(y(k), y(k-1))\mu_2^1(y(k), y(k-1))$$
$$\mu_2(y(k), y(k-1)) = \mu_1^1(y(k), y(k-1))\mu_2^2(y(k), y(k-1))$$
$$\mu_3(y(k), y(k-1)) = \mu_1^2(y(k), y(k-1))\mu_2^1(y(k), y(k-1))$$
$$\mu_4(y(k), y(k-1)) = \mu_1^2(y(k), y(k-1))\mu_2^2(y(k), y(k-1))$$

and fuzzy set A_j^i is characterized by the membership function μ_j^i.

To test the approximating T–S fuzzy system (called *model validation*), its behavior is compared with that of the true robotic link for a variety of inputs. These inputs should be significantly different from the input used to obtain the training data. For instance, if the input current to both the link and the fuzzy system is $u(t) = 10^{-5}$ sign(sin($0.2\pi t$)), the output link angles of Figure 9.7 result. This shows good agreement between the actual link and the T–S fuzzy model. The agreement using a variety of other inputs is also good, therefore the T–S fuzzy system is deemed to approximate the motor-driven robotic link plant to sufficient accuracy.

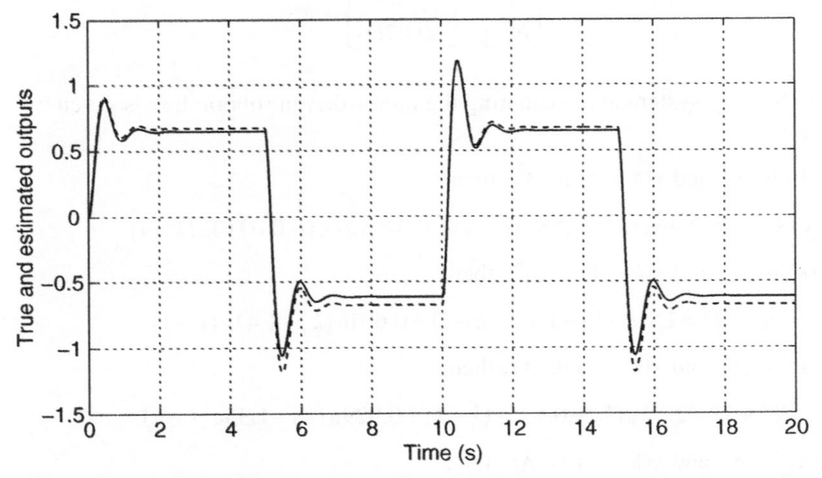

Figure 9.7. Comparison of outputs of true robotic link y (dashed) and approximating fuzzy system $\hat{y}$ (solid), square wave input.

Although it is not the subject of this discussion, we point out that now that we have an accurate T–S model of the link, it is possible to use the methods of 5.3.2 (tracking) and 5.3.3 (model reference) in parallel distributed controllers to control it, similarly to Examples 7.3 and 7.4.

9.2.2 Recursive Least-Squares Identification in Input–Output Difference Equation Form

Recursive least squares can be used to estimate the unknown parameter vector θ according to (8.7). In (8.7a), the correction term for this application should be $\left[y(k+1)-\phi^T(k)\hat{\theta}(k-1)\right]$ with $\phi(k)$ as in (9.7) and

$$\hat{\theta}(k)=\begin{bmatrix}\hat{a}_1^1(k)\\ \vdots\\ \hat{a}_1^R(k)\\ \vdots\\ \hat{a}_n^1(k)\\ \vdots\\ \hat{a}_n^R(k)\\ \hat{b}_1^1(k)\\ \vdots\\ \hat{b}_1^R(k)\\ \vdots\\ \hat{b}_n^1(k)\\ \vdots\\ \hat{b}_n^R(k)\end{bmatrix} \tag{9.9}$$

EXAMPLE 9.4

Consider the system

$$\dot{x}_1=-x_2+x_1^2x_2 \tag{9.10a}$$

$$\dot{x}_2=x_1-x_1^2x_2^3+\left(1+\cos^2 x_1\right)u \tag{9.10b}$$

where $y = x_1$. The system is simulated with a fourth-order Runge–Kutta integration algorithm with a step size of 0.001 s. For the input–output data, we measure pairs $\{u(k\Delta t),\ y(k\Delta t)\}$ for $k = 1, 2, 3, \ldots$ and $\Delta t = 0.01$ s, that is, we measure input and output signals every 10 steps through the RK4 integration routine.

The input function chosen for the identification is a Gaussian pseudorandom sequence with zero mean and a standard deviation of 160. A fuzzy system with $n = 1$ input and $m = 3$ fuzzy sets on its universe is found to be sufficient to accurately approximate the system. Thus there are three rules and six unknown parameters to be determined via RLS. The three input fuzzy sets are chosen as in Figure 9.8.

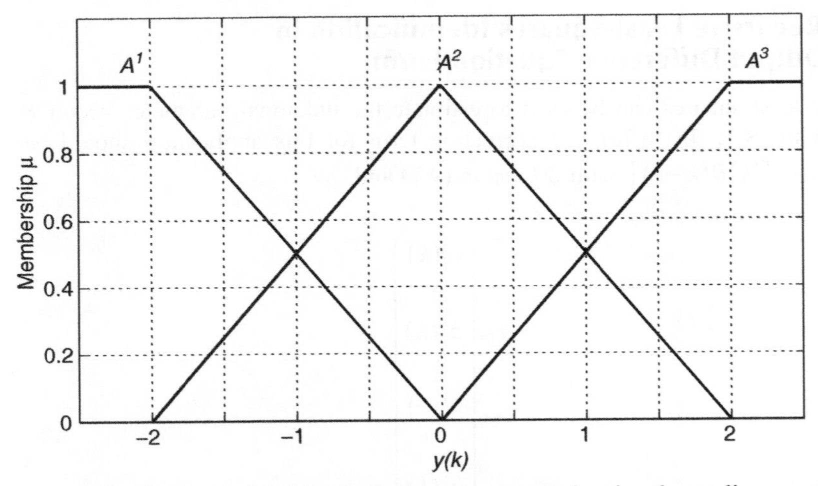

Figure 9.8. Input memberships for fuzzy system approximating the nonlinear system of (9.10).

When the RLS algorithm (8.7) with $\theta(0) = [0\ 0\ 0\ 0\ 0\ 0]^T$ and $P(0) = 10^5 I$ is run simultaneously with this system, the RLS parameter estimates converge after ~5 s (i.e., after ~500 steps) to

$$\theta^* = \begin{bmatrix} a^{1*} \\ a^{2*} \\ a^{3*} \\ b^{1*} \\ b^{2*} \\ b^{3*} \end{bmatrix} = \begin{bmatrix} 0.9926 \\ 1.0077 \\ 0.9901 \\ 2.1176 \times 10^{-5} \\ -1.4328 \times 10^{-5} \\ 1.8814 \times 10^{-5} \end{bmatrix}$$

This results in the following T–S fuzzy rule base:

1. If $y(k)$ is A^1, then $\hat{y}^1(k+1) = 0.9926y(k) + 2.1176 \times 10^{-5}u(k)$.
2. If $y(k)$ is A^2, then $\hat{y}^2(k+1) = 1.0077y(k) - 1.4328 \times 10^{-5}u(k)$.
3. If $y(k)$ is A^3, then $\hat{y}^3(k+1) = 0.9901y(k) + 1.8814 \times 10^{-5}u(k)$.

This rule base together with the input memberships in Figure 9.8 constitute the T–S fuzzy system to closely match the system (9.10). The output of the fuzzy approximator closely matches that of the true system for a variety of inputs, hence the approximator is deemed valid.

Once again, we point out that with the above fuzzy approximation it is now possible to design tracking and model reference controllers for this plant.

9.2.3 Gradient Identification in Input–Output Difference Equation Form

As discussed in Section 8.6, gradient identification does not need linearity in the parameters. Therefore the input fuzzy sets, as well as the output parameters, can be adjusted. Let the fuzzy approximator have R rules of the form:

R_i: If $y(k)$ is A_1^i and $y(k-1)$ is A_2^i and $\cdots$ and $y(k-n+1)$ *is* A_n^i, then

$$y^i(k+1) = a_1^i y(k) + a_2^i y(k-1) + \cdots + a_n^i y(k-n+1) + b_1^i u(k) + b_2^i u(k-1) + \cdots + b_n^i u(k-n+1) \tag{9.11}$$

Notice that these rules are in the form depicted in Figure 8.15, that is, A_j^i is the unique fuzzy set for Rule i, jth universe. The fuzzy system is described by (9.6), where

$$\xi_i(y(k), y(k-1), \cdots, y(k-n+1)) = \frac{\mu_i(y(k), \cdots, y(k-n+1))}{\sum_{i=1}^{R} \mu_i(y(k), \cdots, y(k-n+1))} \tag{9.12}$$

and

$$\mu_i(y(k), \cdots, y(k-n+1)) = \prod_{j=1}^{n} \exp\left(-\frac{1}{2}\left(\frac{y(k-j+1) - c_j^i}{\sigma_j^i}\right)^2\right) \tag{9.13}$$

Let the instantaneous estimation error be defined as

$$e(k) = \frac{1}{2}[\hat{y}(k+1) - y(k+1)]^2 \tag{9.14}$$

We seek to reduce $e(k)$ by choice of parameters $\theta(c_j^i, \sigma_j^i, a_j^i, b_j^i)$. Use $c_j^i(k)$, $\sigma_j^i(k)$, $a_j^i(k)$, $b_j^i(k)$ to denote these parameters' values at time k. By a process similar to that of Sections 8.5 and 8.6, we derive the gradient update laws for these:

$$c_j^i(k+1) = c_j^i(k) - \lambda_1 \varepsilon(k) \left(\frac{\hat{y}^i(k+1) - \hat{y}(k+1)}{\sum_{i=1}^{R} \mu_i(y(k), \cdots, y(k-n+1))}\right) \cdot \mu_i(y(k), \cdots, y(k-n+1)) \left(\frac{y(k-j+1) - c_j^i(k)}{(\sigma_j^i(k))^2}\right) \tag{9.15a}$$

$$\sigma_j^i(k+1) = \sigma_j^i(k) - \lambda_2 \varepsilon(k) \left(\frac{\hat{y}^i(k+1) - \hat{y}(k+1)}{\sum_{i=1}^{R} \mu_i(y(k), \cdots, y(k-n+1))}\right) \cdot \mu_i(y(k), \cdots, y(k-n+1)) \frac{(y(k-j+1) - c_j^i(k))^2}{(\sigma_j^i(k))^3} \tag{9.15b}$$

$$a_j^i(k+1) = a_j^i(k) - \lambda_3 \varepsilon(k) \frac{\mu_i(y(k), \cdots, y(k-n+1))}{\sum_{j=1}^{R} \mu_j(y(k), \cdots, y(k-n+1))} y(k-j+1) \tag{9.15c}$$

$$b_j^i(k+1) = b_j^i(k) - \lambda_4 \varepsilon(k) \frac{\mu_i(y(k), \cdots, y(k-n+1))}{\sum_{j=1}^{R} \mu_j(y(k), \cdots, y(k-n+1))} u(k-j+1) \tag{9.15d}$$

EXAMPLE 9.5

Consider the gantry of Section 1.4, which is a nonminimum-phase nonlinear system. The truth model for the gantry is given by

$$\ddot{\psi} = \frac{-9.81\sin\psi + 1.76\cos\psi\left(-0.18\dot{\psi}^2\sin\psi + u\right)}{0.9 + 0.3\cos^2\psi} \tag{9.16}$$

where ψ is the angle of the rod from vertical-down, $u(t)$ is the input force delivered to the cart, and the output $y = \psi$. The system is simulated with a fourth-order Runge–Kutta integration algorithm with a step size of 0.01 s. For the input–output data, we measure pairs $\{u(k\Delta t), y(k\Delta t)\}$ for $k = 1, 2, 3, \ldots$ and $\Delta t = 0.01$ s.

The input signal chosen for the identification consists of six frequencies covering the operating frequency range of the gantry:

$$u(t) = 1.3(\sin 0.01\pi t + \sin 0.02\pi t + \sin 0.05\pi t + \sin 0.1\pi t + \sin 0.2\pi t + \sin 0.5\pi t)$$

A fuzzy system with1 input and 50 rules is found to be sufficient to accurately approximate the system. Thus there are 200 unknown parameters to be determined via gradient: 50 c's, 50 σ's, 50 a's, and 50 b's. The step sizes chosen by trial and error are $\lambda_1 = \lambda_2 = 5$, $\lambda_3 = \lambda_4 = 0.5$. After 10^5 iterations through the gradient algorithm, the parameters have converged to constant values. The resulting 50-rule T–S fuzzy system produces close agreement with the truth model (9.16) for a variety of inputs. An example is shown in Figure 9.9, where the test input is

$$u(t) = \sin(1.5\pi t) + \sin(0.85\pi t) + \sin(0.15\pi t) \tag{9.17}$$

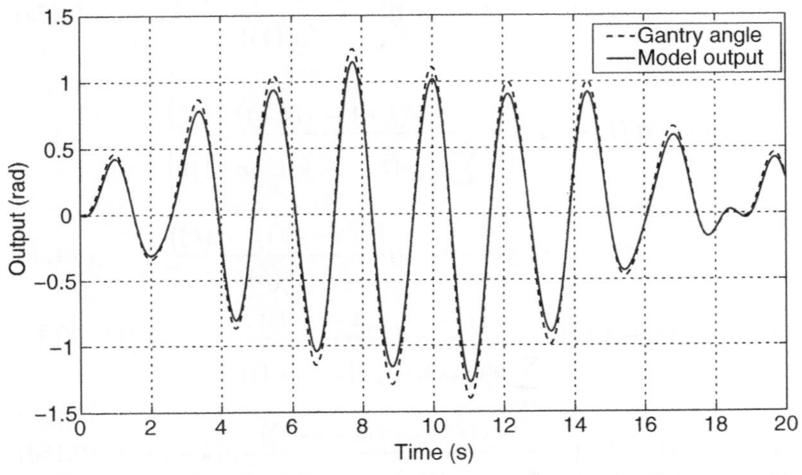

Figure 9.9. Comparison of gantry output and gradient-trained T–S fuzzy model output for test input (9.17).

9.3 IDENTIFICATION IN COMPANION FORM

A continuous-time feedback linearizable system can be modeled in companion form (see Section 5.1.1)

$$y^{(m)} = \delta(x) + \eta(x)u \tag{9.18}$$

where $y^{(m)} = d^m y/dt^m$, x is the vector of states, and m is the relative degree of the system. If the system is unknown, it can be estimated by estimating the nonlinear functions $\delta(x)$ and $\eta(x)$ by fuzzy systems $f_\delta(x)$ and $f_\eta(x)$:

$$y^{(m)} \approx f_\delta(x) + f_\eta(x)u \tag{9.19}$$

9.3.1 Least-Squares Identification in Companion Form

If $f_\delta(x)$ and $f_\eta(x)$ are estimated using least squares (Sections 8.1 or 8.2), the input membership functions must be fixed and the locations of the output singletons are adjustable. Let the rule base of the fuzzy system f_δ consist of R rules of the form

$$R_i: \quad \text{If } x_1 \text{ is } A_1^{K\delta} \text{ and } x_2 \text{ is } A_2^{L\delta} \text{ and} \cdots \text{and } x_n \text{ is } A_n^{M\delta}, \text{ then } f_\delta \text{ is } a^i \tag{9.20}$$

and the rule base of the fuzzy system f_η consist of S rules of the form

$$R_i: \quad \text{If } x_1 \text{ is } A_1^{K\eta} \text{ and } x_2 \text{ is } A_2^{L\eta} \text{ and} \cdots \text{and } x_n \text{ is } A_n^{M\eta}, \text{ then } f_\eta \text{ is } b^i \tag{9.21}$$

Then,

$$f_\delta(x) = \sum_{i=1}^{R} a^i \xi_i \tag{9.22}$$

and

$$f_\eta(x) = \sum_{i=1}^{S} b^i \zeta_i \tag{9.23}$$

where ξ_i, $i = 1, \ldots, R$ are the fuzzy basis functions

$$\xi_i(x) = \frac{\mu_i(x)}{\sum_{i=1}^{R} \mu_i(x)} \tag{9.24}$$

with $\mu_i(x)$ the premise membership function of the ith rule of fuzzy system f_δ, ζ_i, $i = 1, \ldots, S$ are the fuzzy basis functions

$$\zeta_i(x) = \frac{v_i(x)}{\sum_{i=1}^{S} v_i(x)} \tag{9.25}$$

with $v_i(x)$ the premise membership function of the ith rule of fuzzy system f_η, and a^i, $i = 1, \ldots, R$, b^i, $i = 1, \ldots, S$ are unknown parameters to be adjusted by least squares.

Now (9.19) can he written

$$\begin{aligned} y^{(m)} &\approx \sum_{i=1}^{R} a^i \xi_i + \sum_{i=1}^{S} b^i \zeta_i u \\ &= a^1 \xi_1 + \cdots + a^R \xi_R + b^1 \zeta_1 u + \cdots + b^S \zeta_S u \\ &= \phi^T \theta \end{aligned} \tag{9.26}$$

with

$$\phi^T = [\xi_1 \quad \cdots \quad \xi_R \quad \zeta_1 u \quad \cdots \quad \zeta_S u]$$

and

$$\theta = \begin{bmatrix} a^1 & \cdots & a^R & b^1 & \cdots & b^S \end{bmatrix}^T$$

Then, the unknown output parameters a^i and b^i contained in θ can be estimated via batch or recursive least squares as follows.

Let data $\{x(k), u(k), y^{(m)}(k)\}$, $k = 1, 2, 3, \ldots$ be taken from the plant, where $y^{(m)} = d^m y/dt^m$. Define

$$\xi_i(k) = \frac{\mu_i(x(k))}{\sum_{i=1}^{R} \mu_i(x(k))}, i = 1, \ldots, R \tag{9.27}$$

$$\zeta_i(k) = \frac{v_i(x(k))}{\sum_{i=1}^{S} v_i(x(k))}, i = 1, \ldots, S \tag{9.28}$$

and

$$\phi^T(k) = [\xi_1(k) \quad \cdots \quad \xi_R(k) \quad \zeta_1(k)u(k) \quad \cdots \quad \zeta_S(k)u(k)] \tag{9.29}$$

For batch least squares, consider only M of the input–output measurements $\{x(k), u(k), y^{(m)}(k)\}$, $k = 1, \ldots, M$. Form matrices

$$\Phi = \begin{bmatrix} \phi^T(1) \\ \phi^T(2) \\ \vdots \\ \phi^T(M) \end{bmatrix} \sim M \times (R+S)$$

and

$$Y = \begin{bmatrix} y^{(m)}(1) & y^{(m)}(2) & \cdots & y^{(m)}(M) \end{bmatrix}^T \sim M \times 1$$

Then, the batch least-squares estimate of θ based on the M measurements considered is calculated as in (8.6), that is, $\hat{\theta}^* = (\Phi^T \Phi)^{-1} \Phi^T Y$.

For recursive least squares, let the parameter estimate at time k be

$$\hat{\theta}(k) = \begin{bmatrix} \hat{a}^1(k) & \cdots & \hat{a}^R(k) & \hat{b}^1(k) & \cdots & \hat{b}^S(k) \end{bmatrix}^T$$

Then the recursive least-squares algorithm proceeds as in (8.7) with a correction term of $\left[y^{(m)}(k)-\phi^T(k)\hat{\theta}(k-1)\right]$. Therefore, it is necessary not only to know the system's relative degree m, but also to differentiate the system output m times. Although signal differentiation is undesirable in practice, it is unavoidable in this case. For the relative degree, one needs some idea of the system model. In most cases, m is treated as a design parameter and determined by trial and error.

9.3.2 Gradient Identification in Companion Form

If $f_\delta(x)$ and $f_\eta(x)$ are estimated using gradient (Sections 8.5), the input membership functions, as well as the locations of the output singletons, are adjustable. Let the rule base of the fuzzy system f_δ consist of R rules of the form

$$\begin{aligned}
&1.\ \text{If } x_1 \text{ is } A_1^1 \text{ and } x_2 \text{ is } A_2^1 \text{ and} \cdots \text{and } x_n \text{ is } A_n^1, \text{ then } f_\delta \text{ is } a^1.\\
&2.\ \text{If } x_1 \text{ is } A_1^2 \text{ and } x_2 \text{ is } A_2^2 \text{ and} \cdots \text{and } x_n \text{ is } A_n^2, \text{ then } f_\delta \text{ is } a^2.\\
&\vdots\\
&R.\ \text{If } x_1 \text{ is } A_1^R \text{ and } x_2 \text{ is } A_2^R \text{ and} \cdots \text{and } x_n \text{ is } A_n^R, \text{ then } f_\delta \text{ is } a^R.
\end{aligned} \tag{9.30}$$

and the rule base of the fuzzy system f_η consist of S rules of the form

$$\begin{aligned}
&1.\ \text{If } x_1 \text{ is } B_1^1 \text{ and } x_2 \text{ is } B_2^1 \text{ and} \cdots \text{and } x_n \text{ is } B_n^1, \text{ then } f_\eta \text{ is } b^1.\\
&2.\ \text{If } x_1 \text{ is } B_1^2 \text{ and } x_2 \text{ is } B_2^2 \text{ and} \cdots \text{and } x_n \text{ is } B_n^2, \text{ then } f_\eta \text{ is } b^2.\\
&\vdots\\
&S.\ \text{If } x_1 \text{ is } B_1^S \text{ and } x_2 \text{ is } B_2^S \text{ and} \cdots \text{and } x_n \text{ is } B_n^S, \text{ then } f_\eta \text{ is } b^S.
\end{aligned} \tag{9.31}$$

For the rule base of (9.30), assuming we use Gaussian input membership functions, the membership function characterizing A_j^i (ith rule, jth universe) is given by

$$\mu_j^i(x_j)=\exp\left(-\frac{1}{2}\left(\frac{x_j-c_j^i}{\sigma_j^i}\right)^2\right) \tag{9.32}$$

Similarly, the membership function characterizing B_j^i (ith rule, jth universe) is given by

$$v_j^i(x_j)=\exp\left(-\frac{1}{2}\left(\frac{x_j-d_j^i}{\rho_j^i}\right)^2\right) \tag{9.33}$$

Using *product* T-norm, the premise value of Rule i of rule base (9.30) is

$$\mu_i(x_1,x_2,\cdots,x_n)=\prod_{j=1}^{n}\exp\left(-\frac{1}{2}\left(\frac{x_j-c_j^i}{\sigma_j^i}\right)^2\right) \tag{9.34}$$

Then, the fuzzy system f_δ's input–output characteristic is given by

$$f_\delta(x|\theta) = \frac{\sum_{i=1}^{R} a^i \prod_{j=1}^{n} \exp\left(-\frac{1}{2}\left(\frac{x_j - c_j^i}{\sigma_j^i}\right)^2\right)}{\sum_{i=1}^{R} \prod_{j=1}^{n} \exp\left(-\frac{1}{2}\left(\frac{x_j - c_j^i}{\sigma_j^i}\right)^2\right)} = \frac{\sum_{i=1}^{R} a^i \mu_i(x)}{\sum_{i=1}^{R} \mu_i(x)} \tag{9.35}$$

Similarly, the premise value of Rule i of rule base (9.31) is

$$\nu_i(x_1, x_2, \cdots, x_n) = \prod_{j=1}^{n} \exp\left(-\frac{1}{2}\left(\frac{x_j - d_j^i}{\rho_j^i}\right)^2\right) \tag{9.36}$$

Then, the fuzzy system f_η's input–output characteristic is given by

$$f_\eta(x|\theta) = \frac{\sum_{i=1}^{S} b^i \prod_{j=1}^{n} \exp\left(-\frac{1}{2}\left(\frac{x_j - d_j^i}{\rho_j^i}\right)^2\right)}{\sum_{i=1}^{S} \prod_{j=1}^{n} \exp\left(-\frac{1}{2}\left(\frac{x_j - d_j^i}{\rho_j^i}\right)^2\right)} = \frac{\sum_{i=1}^{S} b^i \nu_i(x)}{\sum_{i=1}^{S} \nu_i(x)} \tag{9.37}$$

Now, assume we are given the kth training data triple $\{x(k), u(k), y^{(m)}(k)\}$. Let the instantaneous estimation error corresponding to this triple be defined as

$$e(k) = \frac{1}{2}\left[f_\delta(x(k)|\theta) + f_\eta(x(k)|\theta)u(k) - y^{(m)}(k)\right]^2 \tag{9.38}$$

We seek to reduce $e(k)$ by choice of parameters $\theta\left(a^i, c_j^i, \sigma_j^i, b^i, d_j^i, \rho_j^i\right)$. Use $a^i(k)$, $c_j^i(k)$, $\sigma_j^i(k)$, $b^i(k)$, $d_j^i(k)$, $\rho_j^i(k)$ to denote these parameters' values at time k.

By a process similar to that of Sections 8.5 and 8.6, we derive the gradient update laws for these, where $\varepsilon = f_\delta(x(k)|\theta) + f_\eta(x(k)|\theta)u(k) - y^{(m)}(k)$:

$$a^i(k+1) = a^i(k) - \lambda_1 \varepsilon(k) \frac{\mu_i(x(k))}{\sum_{i=1}^{R} \mu_i(x(k))} \tag{9.39a}$$

$$c_j^i(k+1) = c_j^i(k) - \lambda_2 \varepsilon(k) \left(\frac{a^i(k) - f_\delta(x(k)|\theta(k))}{\sum_{i=1}^{R} \mu_i(x(k))}\right) \mu_i(x(k)) \left(\frac{x_j(k) - c_j^i(k)}{\left(\sigma_j^i(k)\right)^2}\right) \tag{9.39b}$$

$$\sigma_j^i(k+1)=\sigma_j^i(k)-\lambda_3\varepsilon(k)\left(\frac{a^i(k)-f_\delta(x(k)|\theta(k))}{\sum_{i=1}^{R}\mu_i(x(k))}\right)\mu_i(x(k))\frac{(x_j(k)-c_j^i(k))^2}{(\sigma_j^i(k))^3} \quad (9.39c)$$

$$b^i(k+1)=b^i(k)-\lambda_4\varepsilon(k)u(k)\frac{\nu_i(x(k))}{\sum_{i=1}^{S}\nu_i(x(k))} \quad (9.39d)$$

$$d_j^i(k+1)=d_j^i(k)-\lambda_5\varepsilon(k)u(k)\left(\frac{b^i(k)-f_\eta(x(k)|\theta(k))}{\sum_{i=1}^{S}\nu_i(x(k))}\right)\cdot\nu_i(x(k))\left(\frac{x_j(k)-d_j^i(k)}{(\rho_j^i(k))^2}\right) \quad (9.39e)$$

$$\rho_j^i(k+1)=\rho_j^i(k)-\lambda_6\varepsilon(k)u(k)\left(\frac{b^i(k)-f_\eta(x(k)|\theta(k))}{\sum_{i=1}^{S}\nu_i(x(k))}\right)\cdot\nu_i(x(k))\frac{(x_j(k)-d_j^i(k))^2}{(\rho_j^i(k))^3} \quad (9.39f)$$

9.4 SUMMARY

In this chapter, the capabilities of fuzzy systems to estimate static nonlinear functions are utilized for the identification of dynamic plants. This chapter discusses two model forms for the estimated system: **input–output difference equation form** and **companion form**. These are explored because they can he utilized in parallel distributed controllers as in Chapter 7, and in Chapter 10 for adaptive fuzzy control.

If the identification is done in input–output difference equation form, the approximating fuzzy system is of the T–S type with consequents that are input–output difference equations. This form of model can be used for tracking and model reference controllers. If the identification is done in companion form, the fuzzy approximator is a Mamdani fuzzy system whose consequents are unknown constants. This form of model can be used for feedback linearization.

For both of these model forms, we discuss two methods of parameter estimation: **least squares** and **gradient**. These are covered because they are recursive and admit online control. Batch and recursive least-squares methods can only be used for problems in which the system to be estimated is linear in the parameters. Therefore, to be used for the construction of a fuzzy system, the input fuzzy sets must be fixed *a priori* and only the consequent parameters can be adjusted. The gradient method does not require the system to be linear in the parameters, therefore both input and output fuzzy sets can be adjusted.

EXERCISES

9.1 Show that the two-rule fuzzy system of Example 9.1 exactly duplicates the system $\dot{x} = z(x)x$ on the domain $\mathcal{X}$, where $b_m = \min_{\mathcal{X}}(z)$ and $b_M = \max_{\mathcal{X}}(z)$.

9.2 Completely specify a T–S fuzzy system to exactly duplicate the behavior of the forced pendulum (5.4) on the interval $-\pi/4 \le x_1 \le \pi/4$, $-5 \le x_2 \le 5$.

9.3 Find a fuzzy system that is exactly equivalent to the system

$$\dot{x} = x^3 + \cos u$$

on the interval $-2 \le x \le 2$.

9.4 In Example 9.3, assume the number of inputs is 3, the number of fuzzy sets on each input universe is 3, and we have 500 input–output pairs available. (a) How many rules are there? (b) How many unknown parameters are there? (c) Sketch the membership functions characterizing all input fuzzy sets. (d) What is the dimension of Φ? Y?

9.5 In Example 9.3, let a typical rule be

$$R_i: \quad \text{If } y(k) \text{ is } A_1^K \text{ and } y(k-1) \text{ is } A_2^L, \text{ then}$$
$$\hat{y}^i(k+1) = a_1^i y(k) + a_2^i y(k-1) + b_1^i u(k-1) + b_2^i u(k-2)$$

Assume the number of fuzzy sets on each input universe is 2 and we have 200 input–output pairs available. Write the $\phi(k)$ and $\theta(k)$ vectors and give their dimension.

9.6 In Example 9.4, let a typical rule be

$$R_i: \quad \text{If } y(k) \text{ is } A^K, \text{ then } y^i(k+1) = a_1^i y(k) + b_1^i u(k)$$

and each input universe has 4 fuzzy sets. (a) How many rules are there? (b) How many unknown parameters are there? (c) Specify Φ, Y, and θ and give their dimensions. (d) Specify the RLS algorithm including all necessary initial conditions.

9.7 Repeat Problem 9.5 if a typical rule is of the form:

$$R_i: \quad \text{If } y(k) \text{ is } A_1^K \text{ and } y(k-1) \text{ is } A_2^L, \text{ then}$$
$$\hat{y}^i(k+1) = a_1^i y(k) + a_2^i y(k-1) + b_1^i u(k-1) + b_2^i u(k-2)$$

9.8 Consider the gantry of Example 9.5. (a) Is it feedback linearizable? If so, what is its relative degree? (b) Write its model in companion form. (c) Is it controllable? If so, for what ψ? How do you explain this range of ψ for controllability in terms of the physical system?

9.9 Consider the gradient identification in companion form. Fuzzy systems f_δ and f_η each have two inputs and three rules. Write gradient parameter update equations for all adjustable parameters.

9.10 Derive the gradient update laws (9.15) and (9.39).

CHAPTER 10

ADAPTIVE FUZZY CONTROL

An adaptive controller is one that adjusts itself to changes in the plant it is controlling. For instance, consider a robotic arm carrying a load in its end manipulator (hand) and being controlled by an adaptive tracking controller. The controller is designed to force the arm to follow a prescribed reference trajectory through space while moving the load from its start point to its destination. If the mass being moved changes mid-trajectory (say the manipulator accidentally drops some of its load), the controller, because it is adaptive, can adapt to this change and still obtain the same tracking performance as before.

Adaptive controllers fall into two categories [27,33,36–39]: *indirect* and *direct*. The approach in **indirect** adaptive control is to identify a model for the plant, then to derive a controller based on this identified model. The controller calculation is done simultaneously with the plant identification, so model and controller evolve together.

This is called the *certainty equivalence principle*: at each point in the evolution, the controller design is based on the current plant estimate. It is assumed that, as the estimated plant model gets more accurate, the controller also gets more accurate and, in the limit, perfect control is obtained. Furthermore, since the identifier and controller run continuously, if the plant changes in some way (like the robotic arm dropping part of its load in the example above) the estimated model and corresponding controller change correspondingly resulting in a new perfect control.

In **direct** adaptive control, the controller is adjusted directly without referring to any plant model. Thus, the indirect step of identifying the plant is avoided. Instead, the controller adjustment is made on the basis of an error that rates the closed-loop performance.

Plants controlled by adaptive controllers (even nonfuzzy ones) give rise to complex time-varying nonlinear closed-loop systems, whose stability is difficult or impossible to verify. During the last three decades, researchers have met with some success in proving stability of some closed-loop systems containing adaptive controllers. But if the controller is fuzzy, closed-loop stability is much more difficult to prove due to the complexity of fuzzy systems. Therefore, none of the techniques presented in this chapter have been proven to work in all situations, although they may work well in many.

Adaptive controllers run online continuously. Therefore, for indirect adaptive controllers, only recursive parameter estimation algorithms can be used. In Chapters 8 and 9, the only recursive algorithms presented are recursive least-squares (RLS) and gradient. Therefore, we only discuss these two algorithms in connection with indirect adaptive fuzzy control.

10.1 DIRECT ADAPTIVE FUZZY TRACKING CONTROL [12]

Consider the single-input, single-output plant

$$x_p^{(n)} = f_p + g_p u \tag{10.1a}$$

$$y_p = h_p x_p \tag{10.1b}$$

where $x^{(n)} := d^n x/dt^n$, f_p, g_p, and h_p are unknown scalar functions of the plant state $x = [x_p, \dot{x}_p, \cdots, x_p^{(n-1)}]$ and $g_p > 0$ is bounded away from zero. The control objective is to find a plant input, u such that $y_p(t)$ asymptotically tracks a bounded reference signal $r(t)$.

Define the tracking error $e_r = r - y_p$, an error vector $\varepsilon_r = [e_r \quad \dot{e}_r \quad \cdots \quad e_r^{(n-1)}]^T$, and constants $k_1, \ldots, k_n$, such that the roots of $s^n + k_1 s^{n-1} + \ldots + k_{n-1}s + k_n$ are all in the open left half-plane. Define the companion matrix

$$\Lambda = \begin{bmatrix} 0 & 1 & 0 & 0 & \cdots & 0 \\ 0 & 0 & 1 & 0 & \cdots & 0 \\ 0 & 0 & 0 & 1 & \cdots & 0 \\ 0 & 0 & 0 & 0 & \ddots & \vdots \\ \vdots & \vdots & \vdots & \vdots & \cdots & 1 \\ -k_n & -k_{n-1} & -k_{n-2} & -k_{n-3} & \cdots & -k_1 \end{bmatrix} \tag{10.2}$$

Let p_n be the last column of the positive definite matrix P satisfying the Lyapunov matrix equation

$$\Lambda^T P + P\Lambda = -Q \tag{10.3}$$

where Q is a symmetric positive-definite matrix.

Let the inputs to the fuzzy system be $[x_1, x_2, \ldots, x_n] = [x_p, \dot{x}_p, \ldots, x_p^{(n-1)}]$ and define fuzzy sets A_i^j, $i = 1, \ldots, n, j = 1, \ldots, m$ on each universe of discourse. Let the rule base of the direct adaptive fuzzy tracking controller consist of m^n rules of the form

$$R_i\text{:} \qquad \text{If } x_1 \text{ is } A_1^j \text{ and } x_2 \text{ is } A_2^k \text{ and} \cdots \text{and } x_n \text{ is } A_n^l\text{, then } \theta^i \text{ is } \theta_i(t) \tag{10.4}$$

Let $\xi_i(x)$ be the fuzzy basis function for Rule i [in (10.4)] and define $\xi = [\xi_1, \xi_2, \cdots, \xi_{m^n}]$. Defining the parameter vector $\theta = [\theta_1, \theta_2, \cdots \theta_{m^n}]$, the control law is given by

$$u(x|\theta) = \xi(x)\theta(t) \tag{10.5}$$

where the parameter vector $\theta(t)$ is adjusted according to

$$\dot{\theta}(t) = \gamma \varepsilon_r^T p_n \xi(x) \tag{10.6}$$

where $\gamma > 0$ is an arbitrary constant.
Thus the consequents of the fuzzy rules are adjusted online according to (10.6) at each time step.

Note that there is an added supervisory control in [12] guaranteeing boundedness of all signals in the closed-loop system and convergence of $e_r(t)$ to 0 as $t \to \infty$. This is not addressed here. Such guarantees are usually not available for closed-loop systems containing fuzzy controllers, especially adaptive ones.

EXAMPLE 10.1 DIRECT ADAPTIVE FUZZY TRACKING CONTROL FOR BALL AND BEAM

Consider the ball and beam with a time-varying friction (Fig. 1.3) modeled by

$$m\ddot{x} = -\nu x_2 + g\sin(\psi) \tag{10.7}$$

where x is the ball's position on the beam with $x = 0$ corresponding to the middle of the beam, $m = 1$ kg is the ball's mass, ν is the friction encountered by the ball in its travel, $g = 9.81\,\text{m/s}^2$ is the acceleration of gravity, and ψ is the beam angle in radians. Define the plant output $y_p = x$.

The inputs to the direct adaptive fuzzy tracking controller are x and $\dot{x}$. Define three fuzzy sets A_x^1, A_x^2, A_x^3 on the x universe and three fuzzy sets $A_{\dot{x}}^1$, $A_{\dot{x}}^2$, $A_{\dot{x}}^3$ on the $\dot{x}$ universe (see Fig. 10.1). Therefore the fuzzy controller consists of $3^2 = 9$ rules.

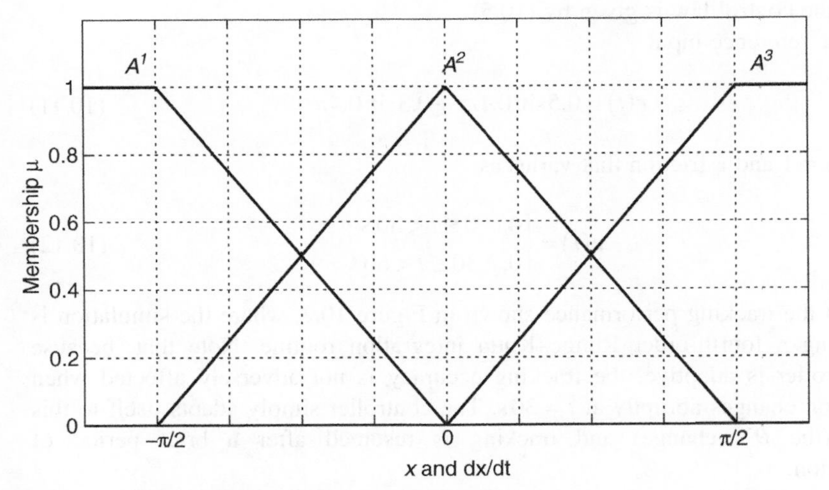

Figure 10.1. Input fuzzy sets on x and $\dot{x}$ universes.

Let $k_1 = 2$ and $k_2 = 1$, so that $s^2 + 2s + 1 = (s + 1)^2$ has both roots in the open left half-plane. Define $Q = \text{diag}(1, 10)$. Then the solution to (10.3) is the positive definite symmetric matrix

$$P = \begin{bmatrix} 3.75 & 0.5 \\ 0.5 & 2.75 \end{bmatrix} \tag{10.8}$$

giving $p_2 = [0.5 \quad 2.75]^T$. The error vector is $\varepsilon_r = [r - y_p \quad \dot{r} - \dot{y}_p]^T$. Setting $\gamma = 10$, the adaptive law (10.6) is given by $\dot{\theta}_i = 10\varepsilon_r^T p_2 \xi_i(x)$, $i = 1, \ldots, 9$ or

$$\dot{\theta}_i = 10\left[0.5(r - y_p) + 2.75(\dot{r} - \dot{y}_p)\right]\xi_i(x), \qquad i = 1, 2, \ldots, 9 \tag{10.9}$$

where the fuzzy basis function $\xi_i(x)$ corresponds to Rule i in the rule base:

$$\begin{aligned}
&1.\ \text{If } x \text{ is } A_x^1 \text{ and } \dot{x} \text{ is } A_{\dot{x}}^1, \text{ then } \theta^1 = \theta_1(t). \\
&2.\ \text{If } x \text{ is } A_x^1 \text{ and } \dot{x} \text{ is } A_{\dot{x}}^2, \text{ then } \theta^2 = \theta_2(t). \\
&3.\ \text{If } x \text{ is } A_x^1 \text{ and } \dot{x} \text{ is } A_{\dot{x}}^3, \text{ then } \theta^3 = \theta_3(t). \\
&4.\ \text{If } x \text{ is } A_x^2 \text{ and } \dot{x} \text{ is } A_{\dot{x}}^1, \text{ then } \theta^4 = \theta_4(t). \\
&5.\ \text{If } x \text{ is } A_x^2 \text{ and } \dot{x} \text{ is } A_{\dot{x}}^2, \text{ then } \theta^5 = \theta_5(t). \\
&6.\ \text{If } x \text{ is } A_x^2 \text{ and } \dot{x} \text{ is } A_{\dot{x}}^3, \text{ then } \theta^6 = \theta_6(t). \\
&7.\ \text{If } x \text{ is } A_x^3 \text{ and } \dot{x} \text{ is } A_{\dot{x}}^1, \text{ then } \theta^7 = \theta_7(t). \\
&8.\ \text{If } x \text{ is } A_x^3 \text{ and } \dot{x} \text{ is } A_{\dot{x}}^2, \text{ then } \theta^8 = \theta_8(t). \\
&9.\ \text{If } x \text{ is } A_x^3 \text{ and } \dot{x} \text{ is } A_{\dot{x}}^3, \text{ then } \theta^9 = \theta_9(t).
\end{aligned} \tag{10.10}$$

Finally, the control law is given by (10.5).

The reference input

$$r(t) = 0.5\sin 0.1\pi t + 0.3\sin 0.4\pi t \tag{10.11}$$

with $x(0) = 1$ and a friction that varies as

$$v(t) = \begin{cases} 4, & 0 \le t < 30\,\text{s} \\ 0, & 30 \le t < 60\,\text{s} \end{cases} \tag{10.12}$$

results in the tracking performance shown in Figure 10.2, where the simulation is done using a fourth-order Runge–Kutta integration routine. Note that, because the controller is adaptive, the tracking accuracy is not adversely affected when the friction changes abruptly at $t = 30$ s. The controller simply adapts itself to this change (the θ's change) and tracking is resumed after a brief period of readaptation.

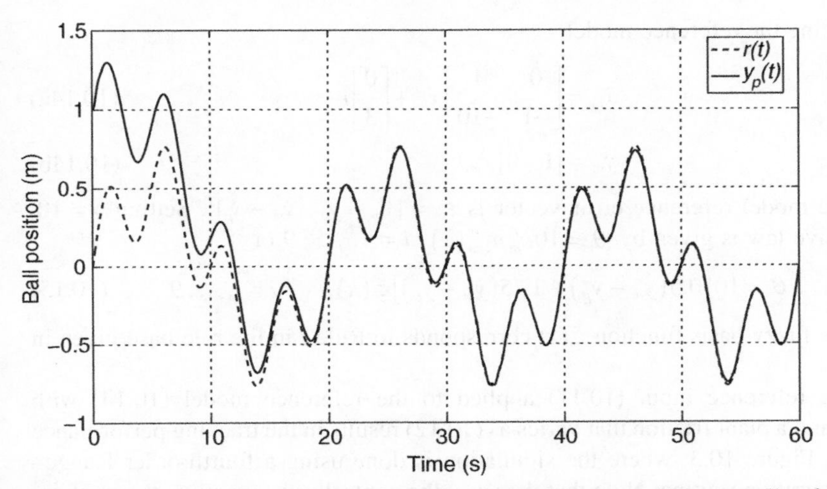

Figure 10.2. Tracking performance of direct adaptive fuzzy tracking controller, friction changing at $t = 30$ s.

10.2 DIRECT ADAPTIVE FUZZY MODEL REFERENCE CONTROL

One version of the model reference control problem is to force the plant output to track the output of a reference model

$$x_m^{(n)} = f_m + g_m r \tag{10.13a}$$

$$y_m = h_m x_m \tag{10.13b}$$

where f_m, g_m, h_m are known scalar functions of the model states $x = \left[x_m, \dot{x}_m, \ldots, x_m^{(n-1)}\right]$ and $r(t)$ is the model input. The derivation of the model reference controller is identical to that of the tracking controller (Section 10.1) except we now define the model following error $e_m = y_m - y_p$ and error vector $\varepsilon_m = \left[e_m, \dot{e}_m, \ldots, e_m^{(n-1)}\right]^T$.

EXAMPLE 10.2 DIRECT ADAPTIVE FUZZY MODEL REFERENCE CONTROL FOR BALL AND BEAM

Consider the ball and beam with a time-varying friction modeled by (10.7), where again $y_p = x$. The inputs to the fuzzy direct adaptive model reference controller are x and $\dot{x}$. Define three fuzzy sets A_x^1, A_x^2, A_x^3 on the x universe and three fuzzy sets $A_{\dot{x}}^1$, $A_{\dot{x}}^2$, $A_{\dot{x}}^3$ on the $\dot{x}$ universe (see Fig. 10.1). Therefore the fuzzy controller has 9 rules. Let $k_1 = 2$, $k_2 = 1$, and $Q = \text{diag}(1, 10)$ as before, again giving $p_2 = [0.5 \quad 2.75]^T$.

Define the reference model

$$\dot{x}_m = \begin{bmatrix} 0 & 1 \\ -1 & -10 \end{bmatrix} x_m + \begin{bmatrix} 0 \\ 3 \end{bmatrix} r \tag{10.14a}$$

$$y_m = \begin{bmatrix} 1 & 0 \end{bmatrix} x_m \tag{10.14b}$$

The model reference error vector is $\varepsilon_m = [y_m - y_p \quad \dot{y}_m - \dot{y}_p]^T$. Setting $\gamma = 10$, the adaptive law is given by $\dot{\theta}_i = 10\varepsilon_m^T p_2 \xi_i(x)$, $i = 1, \ldots, 9$ or

$$\dot{\theta}_i = 10[0.5(y_m - y_p) + 2.75(\dot{y}_m - \dot{y}_p)]\xi_i(x), \qquad i = 1, \ldots, 9 \tag{10.15}$$

where the fuzzy basis function $\xi_i(x)$ corresponds to Rule i in the rule base given in (10.10).

The reference input (10.11) applied to the reference model (10.14) with $x(0) = 1$ and a plant friction that varies as (10.12) results in the tracking performance shown in Figure 10.3, where the simulation is done using a fourth-order Runge–Kutta integration routine. Note that, because the controller is adaptive, the tracking accuracy is not adversely affected when the friction changes abruptly at $t = 30$ s. The controller simply adapts itself to this change (the θ's change) and tracking is resumed after a brief period of readaptation.

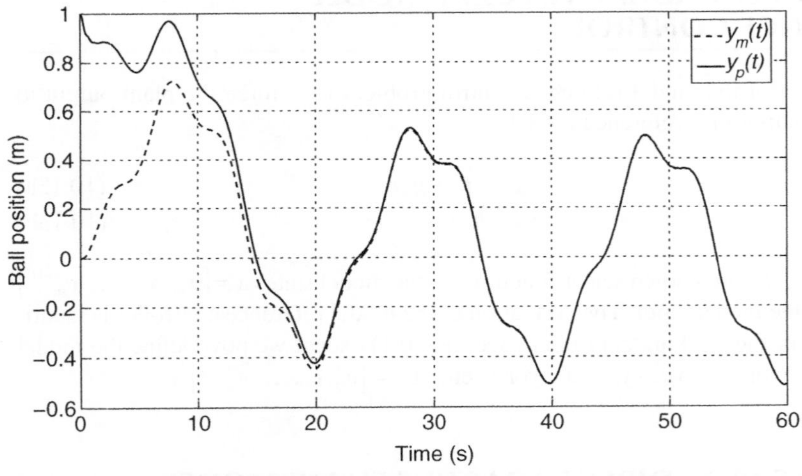

Figure 10.3. Tracking performance of direct adaptive fuzzy model reference controller, friction changing at $t = 30$ s.

Here, we have given one idea for direct adaptive fuzzy control. An interesting idea called *fuzzy model reference learning control* (FMRLC) is another well-known type of direct adaptive fuzzy control. The FMRLC goes beyond what is presented here in that the controller not only learns but remembers what it has learned. Thus it is slightly more involved than what we have presented here. The interested reader is referred to [9,40] for more information on FMRLC.

10.3 INDIRECT ADAPTIVE FUZZY TRACKING CONTROL

As discussed in Sections 9.2.2 and 9.2.3, recursive least squares or gradient can be used to recursively identify a T–S fuzzy model for any time-invariant system. The T–S fuzzy system (called the *plant fuzzy system*) has rules of the form (9.4), that is, the rule consequents are input–output difference equations. At each time step, the parameter estimates in the consequents change and eventually converge to constant values.

Section 7.3 shows how to design a parallel distributed tracking controller for a time-invariant system modeled as a T–S fuzzy system with rules of the form (9.4). Using the certainty equivalence principle at each time step, we can employ this design method to derive a parallel distributed tracking controller (called the *controller fuzzy system*). The parameters of the controller fuzzy system change at each time step in response to the changing parameters in the plant fuzzy system. If the identification–control design process described above is left running online, the controller can adapt to any change in the plant being controlled and still produce accurate tracking even in the presence of such change.

In contrast to direct adaptive control, the indirect adaptive control strategy takes the additional step of identifying a model for the plant. In fact, this is one of the advantages of the indirect control strategy: A plant model is obtained as a byproduct of the control method. Note, however, that we are not as interested in accurately identifying the plant as in controlling it, although acquiring an accurate plant model is certainly a valuable byproduct.

EXAMPLE 10.3 INDIRECT ADAPTIVE FUZZY TRACKING CONTROL FOR GANTRY (RLS PARAMETER ADAPTATION)

Consider the gantry of Example 9.5 whose truth model (assumed unknown) is given by (9.16), where ψ is the angle of the rod from vertical-down, $u(t)$ is the input force delivered to the cart, and the output $y = \psi$. It is desired to formulate a force input to the cart so that the pendulum angle tracks the reference signal

$$r(t) = 0.2\left(\sin 0.5\pi t + \sin 0.7\pi t + \sin\sqrt{2}\pi t\right) \qquad (10.16)$$

Since the gantry model is unknown, we use recursive least-squares parameter estimation to identify it together with a certainty-equivalent parallel distributed adaptive tracking control strategy.

Through trial and error, it is determined that a Takagi–Sugeno (T–S) fuzzy system with two inputs ($y(k)$ and $y(k-1)$) and four fuzzy sets characterized by triangular membership functions forming partitions of unity on $[-\pi/2, \pi/2]$ on each universe (A_j^i, $j = 1, 2$, $i = 1, 2, 3, 4$) is sufficient to accurately control the system.

The memberships characterizing the input fuzzy sets are shown in Figure 10.4. Therefore, there are 16 rules and 64 parameters to be identified.

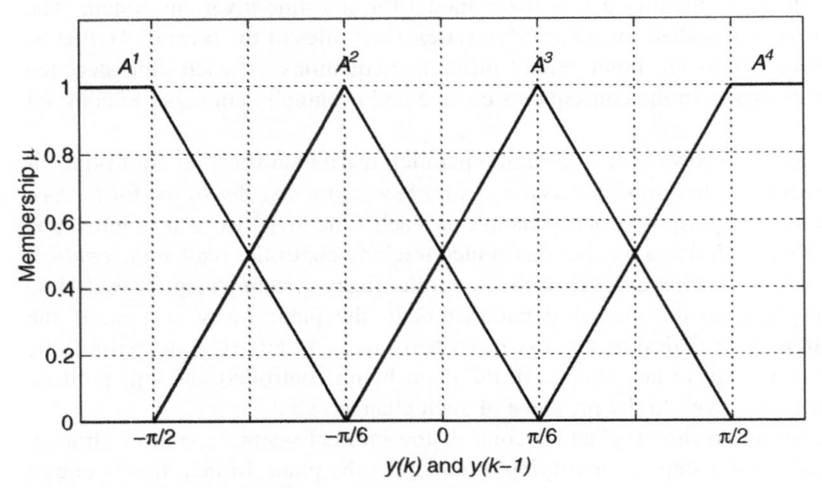

Figure 10.4. Input fuzzy sets on $y(k)$ and $y(k-1)$ universes.

For the RLS algorithm initial conditions, we set $P(0) = 10^5 I_{64}$ and $\theta_i(0) = 0.001$, $i = 1, \ldots, 64$. The step size of the RK4 routine used for the simulation is 0.01 s. A typical rule in the rule base of the plant fuzzy system is

R_{ip}: If $y(k)$ is A_1^i and $y(k-1)$ is A_2^i, then

$$\hat{y}^i(k+1) = \hat{a}_1^i y(k) + \hat{a}_2^i y(k-1) + \hat{b}_1^i u(k) + \hat{b}_2^i u(k-1)$$

and a typical rule in the rule base of the controller fuzzy system is

R_{ic}: If $y(k)$ is A_1^i and $y(k-1)$ is A_2^i, then

$$u^i(k) = \frac{1}{\hat{b}_1^i(k)}\left[-\hat{b}_2^i u(k-1) - \hat{a}_1^i y(k) - \hat{a}_2^i y(k-1) + r(k+1)\right]$$

Because of the $1/\hat{b}_1^i(k)$ term in the consequent of the controller fuzzy system, it is necessary to insure that $\hat{b}_1^i(k)$ is bounded away from zero for all k to avoid division by zero.

The tracking performance of the closed-loop system consisting of the gantry and the indirect adaptive tracking controller is shown in Figure 10.5. After an adaptation period of ~8 s, the rod angle tracks the reference signal well.

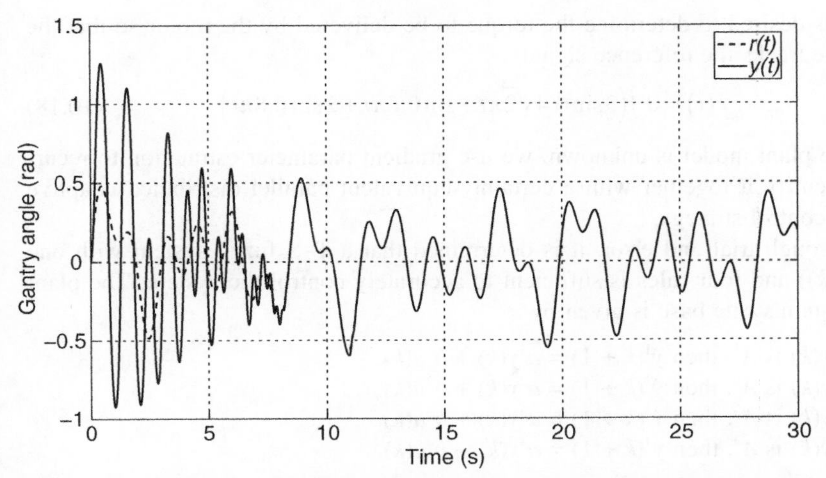

Figure 10.5. Tracking performance of indirect adaptive fuzzy tracking controller for gantry, RLS parameter adaptation.

EXAMPLE 10.4 INDIRECT ADAPTIVE FUZZY TRACKING CONTROL FOR MOTOR-DRIVEN ROBOTIC LINK (GRADIENT PARAMETER ADAPTATION)

Consider the motor-driven robotic link shown in Figure 10.6, whose truth model (assumed unknown) is given by

$$I\ddot{\psi} = -mgl\cos\psi - B\dot{\psi} + \tau \tag{10.17}$$

where ψ is the link angle from horizontal in radians, $m = 1\,\text{kg}$ is the mass of the link, $l = 0.25\,\text{m}$ is the distance from the motor shaft to the link's center of mass, $I = ml^2$ (kg-m^2) is the rotational inertia of the shaft and load, $B = 0.7\,\text{kg-m}^2/\text{s}$ is the rotational friction of the motor shaft, $g = 9.81\,\text{m/s}^2$ is the acceleration of gravity, and τ (kg-m^2/s^2) is the torque delivered to the link by the motor. The output is the measured link angle $y(t) = \psi(t)$.

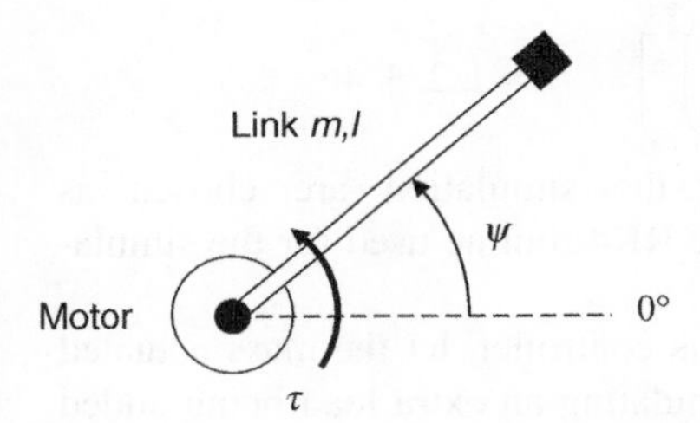

Figure 10.6. Robotic link of Example 10.5.

It is desired to determine the torque to be delivered by the motor so that the link angle tracks the reference signal

$$r(t)=0.1\left(3\sin 0.4\sqrt{2}\pi t+\sin 0.6\pi t+2\sin 0.8\pi t\right) \tag{10.18}$$

Since the plant model is unknown, we use gradient parameter estimation to recursively identify it together with a certainty-equivalent parallel distributed adaptive tracking control strategy.

Through trial and error, it is determined that a T–S fuzzy system with one input ($y(k)$) and four rules is sufficient to accurately control the system. The plant fuzzy system's rule base is given by

1. If $y(k)$ is A^1, then $\hat{y}^1(k+1) = a^1 y(k) + b^1 u(k)$.
2. If $y(k)$ is A^2, then $\hat{y}^2(k+1) = a^2 y(k) + b^2 u(k)$.
3. If $y(k)$ is A^3, then $\hat{y}^3(k+1) = a^3 y(k) + b^3 u(k)$.
4. If $y(k)$ is A^4, then $\hat{y}^4(k+1) = a^4 y(k) + b^4 u(k)$.

where input fuzzy sets A^1, A^2, A^3, and A^4 are characterized by Gaussian membership functions whose centers and spreads are to be determined via gradient parameter adaptation. In all, there are 16 parameters to be identified: four input membership function centers (c^i, $i = 1, 2, 3, 4$), four input membership spreads (σ^i, $i = 1, 2, 3, 4$), and eight consequent parameters (a^i and b^i, $i = 1, 2, 3, 4$). These are adjusted as in (9.15), giving $\hat{c}^i(k)$, $\hat{\sigma}^i(k)$, $\hat{a}^i(k)$, $\hat{b}^i(k)$, $i = 1, 2, 3, 4$ and all integer k.

At each time increment, the certainty equivalent tracking controller fuzzy system is given by

1. If $y(k)$ is A^1, then $u^1(k) = (1/\hat{b}^1)(-\hat{a}^1 y(k) + r(k+1))$.
2. If $y(k)$ is A^2, then $u^2(k) = (1/\hat{b}^2)(-\hat{a}^2 y(k) + r(k+1))$.
3. If $y(k)$ is A^3, then $u^3(k) = (1/\hat{b}^3)(-\hat{a}^3 y(k) + r(k+1))$.
4. If $y(k)$ is A^4, then $u^4(k) = (1/\hat{b}^4)(-\hat{a}^4 y(k) + r(k+1))$.

The resulting torque delivered to the link at time k is

$$u(k)=\xi_1(k)u^1(k)+\xi_2(k)u^2(k)+\xi_3(k)u^3(k)+\xi_4(k)u^4(k)$$

where the fuzzy basis functions are given by

$$\xi_i(k)=\frac{\mu_i(y(k))}{\mu_1(y(k))+\mu_2(y(k))+\mu_3(y(k))+\mu_4(y(k))} \qquad i=1,2,3,4$$

and the premise value of Rule i is given by

$$\mu_i(y(k))=\exp\left(-\frac{1}{2}\left(\frac{y(k)-\hat{c}^i(k)}{\hat{\sigma}^i(k)}\right)^2\right) \qquad i=1,2,3,4$$

The gradient algorithm step sizes for the simulation are chosen as $\lambda_1 = \lambda_2 = \lambda_3 = \lambda_4 = 3 \times 10^{-2}$. The step size of the RK4 routine used for the simulation is 0.01 s.

To illustrate the adaptive capabilities of this controller, let the mass actuated by the link change from 1 to 5 kg at $t = 30$ s, simulating an extra load being added to the link. The tracking performance of the closed-loop system consisting of the robotic link and the indirect adaptive tracking controller is shown in Figure 10.7.

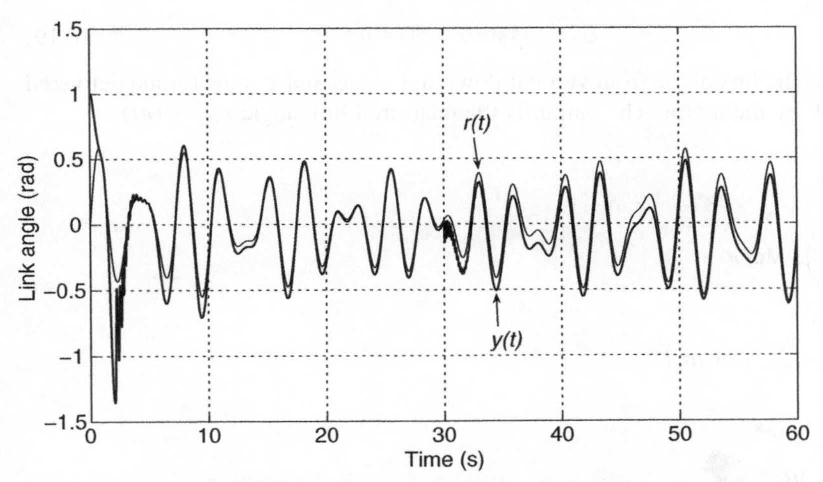

Figure 10.7. Tracking the performance of a robotic link controlled by an indirect adaptive fuzzy controller with gradient parameter adaptation, link mass increasing fivefold at $t = 30$ s.

After an adaptation period of ~20 s, the rod angle tracks the reference signal well. Note that increasing the load fivefold at $t = 30$ s does result in a slight degradation in tracking performance. Nevertheless, approximate tracking is retained even after such a drastic increase in load mass. This is because the adaptive nature of the controller allows it to adjust itself to the new operating conditions.

10.4 INDIRECT ADAPTIVE FUZZY MODEL REFERENCE CONTROL

Assume the plant is modeled recursively as a T–S fuzzy system with rules of the form (9.4). Section 7.4 shows how to design a parallel distributed model reference controller for such a T–S fuzzy system. Using the certainty equivalence principle at each time step, we can employ this design technique to derive a parallel distributed model reference controller for the plant. The parameters of the controller fuzzy system change at each time step in response to the changing parameters in the plant fuzzy system.

If the identification–control design process described above is left running online, the controller can adapt to any change in the plant being controlled and still produce accurate tracking even in the presence of such change.

EXAMPLE 10.5 INDIRECT ADAPTIVE FUZZY MODEL REFERENCE CONTROL FOR MOTOR-DRIVEN ROBOTIC LINK (RLS PARAMETER ADAPTATION)

Consider the motor-driven robotic link shown in Figure 10.8, whose truth model (assumed unknown) is given by

$$\ddot{\psi} = -64\sin\psi - 5\dot{\psi} + 4\tau \tag{10.19}$$

where ψ is the link angle from vertical-down in radians and τ is the torque delivered to the link by the motor. The output is the measured link angle $y(t) = \psi(t)$.

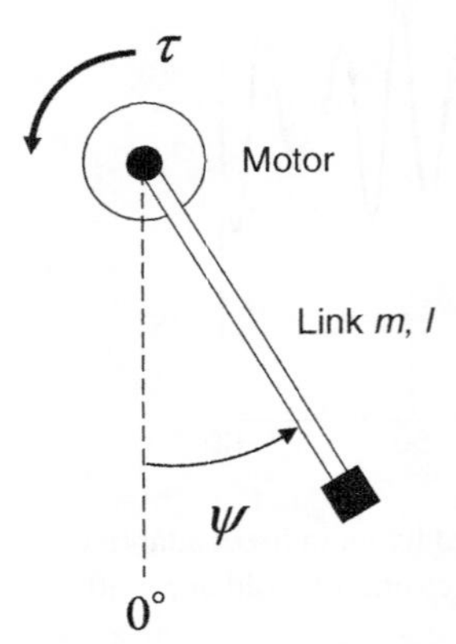

Figure 10.8. Robotic link of Example 10.5.

It is desired that the robotic link plant follow a model with transfer function

$$H(z) = \frac{z - 0.8}{z^2 - 1.9z + 0.9125} \tag{10.20}$$

Thus in the notation of Section 5.3.3 we have $g = 1$, $H(q^{-1}) = 1 - 0.8q^{-1}$, and $E(q^{-1}) = 1 - 1.9q^{-1} + 0.9125q^{-2}$. Therefore, the reference model can be characterized by the difference equation

$$y_m(k+1) = 1.9y_m(k) - 0.9125y_m(k-1) + r(k) - 0.8r(k-1) \tag{10.21}$$

where $r(k)$ is an arbitrary reference input.

Through trial and error, it is determined that a T–S fuzzy system with two inputs [$y(k)$ and $y(k - 1)$] and three fuzzy sets characterized by triangular membership functions forming partitions of unity on $[-\pi/2, \pi/2]$ on each universe (A_j^i , $j = 1, 2$, $i = 1, 2, 3$) is sufficient to accurately control the system. The memberships characterizing the input fuzzy sets are shown in Figure 10.9. Therefore, there are 9 rules and 36 parameters to be identified.

The plant fuzzy system's rule base is given by

1. If $y(k)$ is A_1^1 and $y(k-1)$ is A_2^1, then

$$\hat{y}^1(k+1) = a_1^1 y(k) + a_2^1 y(k-1) + b_1^1 u(k) + b_2^1 u(k-1)$$

2. If $y(k)$ is A_1^1 and $y(k-1)$ is A_2^2, then

$$\hat{y}^2(k+1) = a_1^2 y(k) + a_2^2 y(k-1) + b_1^2 u(k) + b_2^2 u(k-1)$$

3. If $y(k)$ is A_1^1 and $y(k-1)$ is A_2^3, then

$$\hat{y}^3(k+1) = a_1^3 y(k) + a_2^3 y(k-1) + b_1^3 u(k) + b_2^3 u(k-1)$$

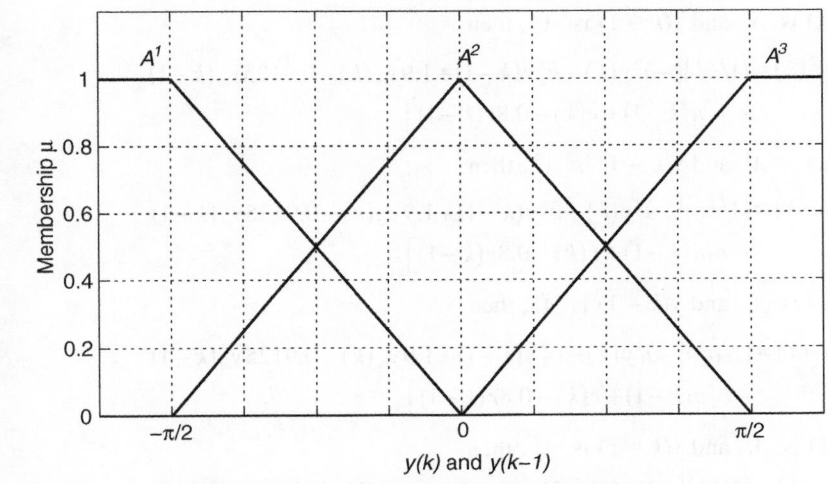

Figure 10.9. Input fuzzy sets on $y(k)$ and $y(k - 1)$ universes.

4. If $y(k)$ is A_1^2 and $y(k - 1)$ is A_2^1, then

$$\hat{y}^4(k+1) = a_1^4 y(k) + a_2^4 y(k-1) + b_1^4 u(k) + b_2^4 u(k-1)$$

5. If $y(k)$ is A_1^2 and $y(k - 1)$ is A_2^2, then

$$\hat{y}^5(k+1) = a_1^5 y(k) + a_2^5 y(k-1) + b_1^5 u(k) + b_2^5 u(k-1)$$

6. If $y(k)$ is A_1^2 and $y(k - 1)$ is A_2^3, then

$$\hat{y}^6(k+1) = a_1^6 y(k) + a_2^6 y(k-1) + b_1^6 u(k) + b_2^6 u(k-1)$$

7. If $y(k)$ is A_1^3 and $y(k - 1)$ is A_2^1, then

$$\hat{y}^7(k+1) = a_1^7 y(k) + a_2^7 y(k-1) + b_1^7 u(k) + b_2^7 u(k-1)$$

8. If $y(k)$ is A_1^3 and $y(k - 1)$ is A_2^2, then

$$\hat{y}^8(k+1) = a_1^8 y(k) + a_2^8 y(k-1) + b_1^8 u(k) + b_2^8 u(k-1)$$

9. If $y(k)$ is A_1^3 and $y(k - 1)$ is A_2^3, then

$$\hat{y}^9(k+1) = a_1^9 y(k) + a_2^9 y(k-1) + b_1^9 u(k) + b_2^9 u(k-1)$$

where $u = \tau$, the torque delivered to the link by the motor. The 36 parameters to be identified: a_j^i and b_j^i, $i = 1, \ldots 9$, $j = 1, 2$ are adjusted as in (8.7), giving $\hat{a}_j^i(k)$, $\hat{b}_j^i(k)$, $i = 1, \ldots 9$, $j = 1, 2$, and $k = 1, 2, 3, \ldots$.

At each time increment, the certainty equivalent model reference controller fuzzy system is given by

1. If $y(k)$ is A_1^1 and $y(k - 1)$ is A_2^1, then

$$u^1(k) = \left(1/\hat{b}_1^1\right)\left[-\hat{a}_1^1 y(k) - \hat{a}_2^1 y(k-1) + 1.9y_m(k) - 0.9125y_m(k-1) - \hat{b}_2^1 u(k-1) + r(k) - 0.8r(k-1)\right]$$

2. If $y(k)$ is A_1^1 and $y(k-1)$ is A_2^2, then

$$u^2(k) = \left(1/\hat{b}_1^2\right)\left[-\hat{a}_1^2 y(k) - \hat{a}_2^2 y(k-1) + 1.9y_m(k) - 0.9125y_m(k-1) - \hat{b}_2^2 u(k-1) + r(k) - 0.8r(k-1)\right]$$

3. If $y(k)$ is A_1^1 and $y(k-1)$ is A_2^3, then

$$u^3(k) = \left(1/\hat{b}_1^3\right)\left[-\hat{a}_1^3 y(k) - \hat{a}_2^3 y(k-1) + 1.9y_m(k) - 0.9125y_m(k-1) - \hat{b}_2^3 u(k-1) + r(k) - 0.8r(k-1)\right]$$

4. If $y(k)$ is A_1^2 and $y(k-1)$ is A_2^1, then

$$u^4(k) = \left(1/\hat{b}_1^4\right)\left[-\hat{a}_1^4 y(k) - \hat{a}_2^4 y(k-1) + 1.9y_m(k) - 0.9125y_m(k-1) - \hat{b}_2^4 u(k-1) + r(k) - 0.8r(k-1)\right]$$

5. If $y(k)$ is A_1^2 and $y(k-1)$ is A_2^2, then

$$u^5(k) = \left(1/\hat{b}_1^5\right)\left[-\hat{a}_1^5 y(k) - \hat{a}_2^5 y(k-1) + 1.9y_m(k) - 0.9125y_m(k-1) - \hat{b}_2^5 u(k-1) + r(k) - 0.8r(k-1)\right]$$

6. If $y(k)$ is A_1^2 and $y(k-1)$ is A_2^3, then

$$u^6(k) = \left(1/\hat{b}_1^6\right)\left[-\hat{a}_1^6 y(k) - \hat{a}_2^6 y(k-1) + 1.9y_m(k) - 0.9125y_m(k-1) - \hat{b}_2^6 u(k-1) + r(k) - 0.8r(k-1)\right]$$

7. If $y(k)$ is A_1^3 and $y(k-1)$ is A_2^1, then

$$u^7(k) = \left(1/\hat{b}_1^7\right)\left[-\hat{a}_1^7 y(k) - \hat{a}_2^7 y(k-1) + 1.9y_m(k) - 0.9125y_m(k-1) - \hat{b}_2^7 u(k-1) + r(k) - 0.8r(k-1)\right]$$

8. If $y(k)$ is A_1^3 and $y(k-1)$ is A_2^2, then

$$u^8(k) = \left(1/\hat{b}_1^8\right)\left[-\hat{a}_1^8 y(k) - \hat{a}_2^8 y(k-1) + 1.9y_m(k) - 0.9125y_m(k-1) - \hat{b}_2^8 u(k-1) + r(k) - 0.8r(k-1)\right]$$

9. If $y(k)$ is A_1^3 and $y(k-1)$ is A_2^3, then

$$u^9(k) = \left(1/\hat{b}_1^9\right)\left[-\hat{a}_1^9 y(k) - \hat{a}_2^9 y(k-1) + 1.9y_m(k) - 0.9125y_m(k-1) - \hat{b}_2^9 u(k-1) + r(k) - 0.8r(k-1)\right]$$

The resulting torque delivered to the link by the motor at time k is

$$\tau(k) = u(k) = \sum_{i=1}^{9} \xi_i(k) u^i(k)$$

where the fuzzy basis functions are given by

$$\xi_i(k) = \frac{\mu_i(y(k))}{\sum_{j=1}^{9} \mu_j(y(k))}, \qquad i = 1, \ldots, 9$$

Let the signal delivered to the reference model (10.21) be

$$r(t) = 0.6\left(0.1\sin 0.4\sqrt{2}\pi t + 0.8\sin \pi t + 0.5\sin 0.2\pi t\right) \tag{10.22}$$

The model following performance of the closed-loop system consisting of the robotic link and the indirect adaptive model reference controller is shown in Figure 10.10. After an adaptation period of ~2 s, the rod angle tracks the reference model well.

To further test the controller, the refererence model is excited by the signal

$$r(t) = 0.75\,\mathrm{sign}(\sin 0.2\pi t) \tag{10.23}$$

All controller parameters remain unchanged from the previous run (Fig. 10.10). The resulting comparison between the plant and model outputs is shown in Figure 10.11. Again, the agreement is quite good.

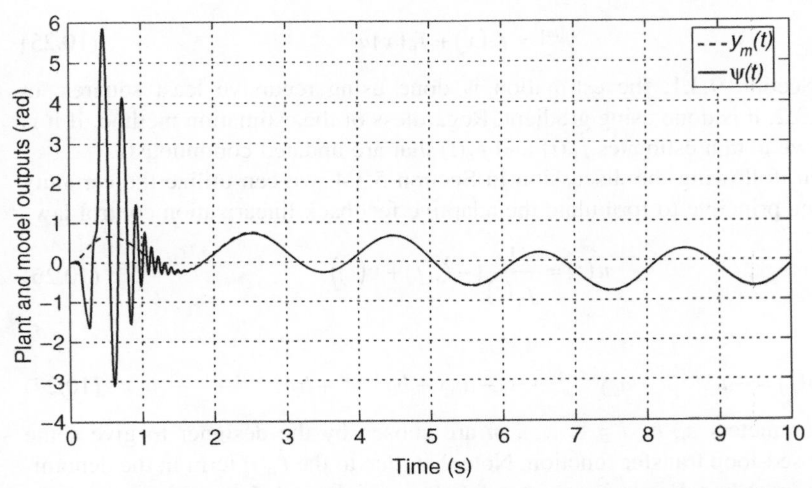

Figure 10.10. Model following performance of robotic link controlled by indirect adaptive fuzzy controller with RLS parameter adaptation. Reference model input given by (10.22).

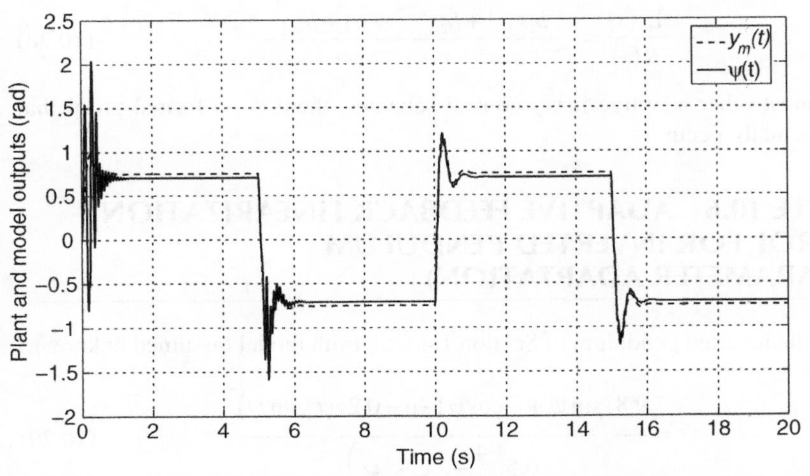

Figure 10.11. Model following performance of robotic link controlled by indirect adaptive fuzzy controller with RLS parameter adaptation. Reference model input given by (10.23).

10.5 ADAPTIVE FEEDBACK LINEARIZATION CONTROL

As discussed in Section 9.3, a continuous-time feedback linearizable system can be modeled in companion form

$$y^{(m)} = \delta(x) + \eta(x)u \tag{10.24}$$

where u is the scalar input, y is the scalar output, $y^{(m)} = d^m y/dt^m$, x is the vector of states, and m is the relative degree of the system. If the system is unknown, it can be estimated by estimating the nonlinear functions $\delta(x)$ and $\eta(x)$ by fuzzy systems $f_\delta(x)$ and $f_\eta(x)$:

$$y^{(m)} \approx f_\delta(x) + f_\eta(x)u \tag{10.25}$$

In Section 9.3.1, the estimation is done using recursive least squares; in Section 9.3.2, it is done using gradient. Regardless of the estimation method, if it is recursive we obtain estimates $\hat{f}_\delta(t)$ and $\hat{f}_\eta(t)$ that are updated continuously.

Then, following the discussion in Section 5.3.4, we can utilize the certainty equivalence principle to formulate the adaptive feedback linearization control law

$$u(t) = \frac{1}{\hat{f}_\eta(t)}\left(-\hat{f}_\delta(t) + v(t)\right) \tag{10.26}$$

where

$$v(t) = -a_1 y^{(m-1)} - a_2 y^{(m-2)} - \cdots - a_m y + b_1 r^{(m-1)} + b_2 r^{(m-2)} + \cdots + b_m r \tag{10.27}$$

and the parameters a_i, b_i, $i = 1, \ldots, m$ are chosen by the designer to give some desired closed-loop transfer function. Note that, due to the $\hat{f}_\eta(t)$ term in the denominator, care must be taken to insure that $\hat{f}_\eta(t)$ remains bounded away from zero.

The control objective is that the closed-loop system asymptotically approaches a linear system with transfer function

$$\frac{Y_m(s)}{R(s)} = \frac{b_1 s^{m-1} + b_2 s^{m-2} + \cdots + b_m}{s^m + a_1 s^{m-1} + a_2 s^{m-2} + \cdots + a_m} \tag{10.28}$$

As with most other adaptive fuzzy control schemes, there is no formal proof that this will actually occur.

EXAMPLE 10.6 ADAPTIVE FEEDBACK LINEARIZATION CONTROL FOR INVERTED PENDULUM (RLS PARAMETER ADAPTATION)

Consider the inverted pendulum of Section 1.4 with truth model (assumed unknown)

$$\ddot{\psi} = \frac{9.81\sin\psi + \frac{2}{3}\cos\psi\left(-u - 0.25\dot{\psi}^2 \sin\psi\right)}{0.5\left(\frac{4}{3} - \frac{1}{3}\cos^2\psi\right)} \tag{10.29}$$

where $\psi(t)$ is the pendulum angle from the vertical-up position and u is the force delivered to the cart. It can be shown that this plant is feedback linearizable with relative degree $m = 2$. Therefore, it can be modeled in companion form (10.24) with $\delta(\psi, \dot{\psi})$ and $\eta(\psi, \dot{\psi})$ unknown nonlinear functions of the pendulum angle ψ and its derivative.

We can use RLS to recursively approximate the functions $\delta(\psi, \dot{\psi})$ and $\eta(\psi, \dot{\psi})$ with fuzzy systems, as discussed in Section 9.3.1. Define three fuzzy sets on the ψ universe characterized by three triangular memberships shown in Figure 10.12 and three fuzzy sets on the $\dot{\psi}$ universe characterized by three triangular memberships shown in Figure 10.13.

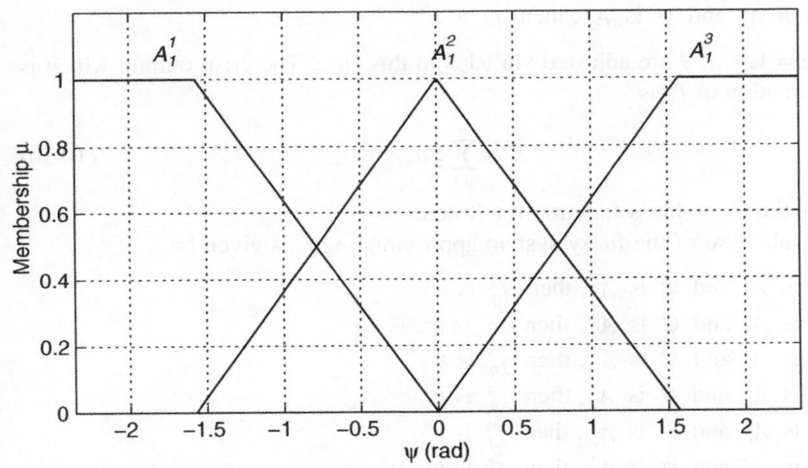

Figure 10.12. Membership functions characterizing input fuzzy sets on ψ universe, RLS approximation of inverted pendulum plant (10.29).

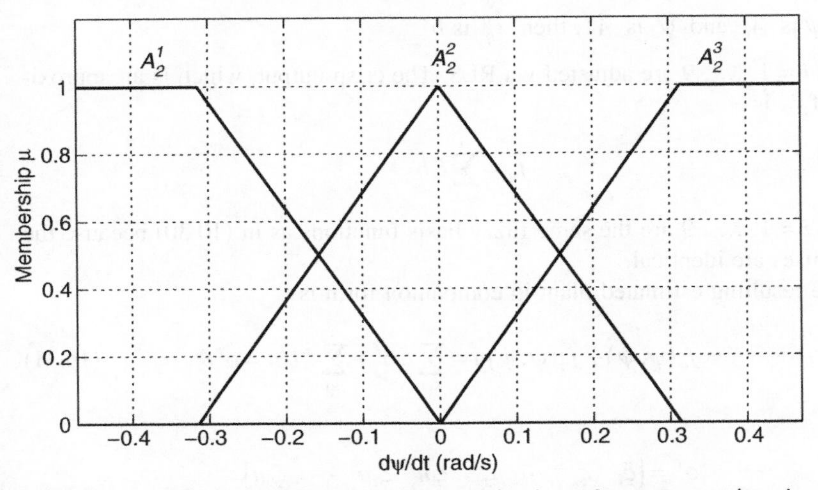

Figure 10.13. Membership functions characterizing input fuzzy sets on $\dot{\psi}$ universe, RLS approximation of inverted pendulum plant (10.29).

The rule base of the fuzzy system approximating f_δ is given by

1. If ψ is A_1^1 and $\dot{\psi}$ is A_2^1, then $\hat{f}_\delta^1$ is a^1.
2. If ψ is A_1^1 and $\dot{\psi}$ is A_2^2, then $\hat{f}_\delta^2$ is a^2.
3. If ψ is A_1^1 and $\dot{\psi}$ is A_2^3, then $\hat{f}_\delta^3$ is a^3.
4. If ψ is A_1^2 and $\dot{\psi}$ is A_2^1, then $\hat{f}_\delta^4$ is a^4.
5. If ψ is A_1^2 and $\dot{\psi}$ is A_2^2, then $\hat{f}_\delta^5$ is a^5.
6. If ψ is A_1^2 and $\dot{\psi}$ is A_2^3, then $\hat{f}_\delta^6$ is a^6.
7. If ψ is A_1^3 and $\dot{\psi}$ is A_2^1, then $\hat{f}_\delta^7$ is a^7.
8. If ψ is A_1^3 and $\dot{\psi}$ is A_2^2, then $\hat{f}_\delta^8$ is a^8.
9. If ψ is A_1^3 and $\dot{\psi}$ is A_2^3, then $\hat{f}_\delta^9$ is a^9.

where a^i, $i = 1, \ldots, 9$ are adjusted via RLS in this case. The crisp output, which is an approximation of f_δ, is

$$\hat{f}_\delta = \sum_{i=1}^{9} \xi_i a^i \tag{10.30}$$

where ξ_i is the fuzzy basis function for Rule i.

The rule base of the fuzzy system approximating f_η is given by

1. If ψ is A_1^1 and $\dot{\psi}$ is A_2^1, then $\hat{f}_\eta^1$ is b^1.
2. If ψ is A_1^1 and $\dot{\psi}$ is A_2^2, then $\hat{f}_\eta^2$ is b^2.
3. If ψ is A_1^1 and $\dot{\psi}$ is A_2^3, then $\hat{f}_\eta^3$ is b^3.
4. If ψ is A_1^2 and $\dot{\psi}$ is A_2^1, then $\hat{f}_\eta^4$ is b^4.
5. If ψ is A_1^2 and $\dot{\psi}$ is A_2^2, then $\hat{f}_\eta^5$ is b^5.
6. If ψ is A_1^2 and $\dot{\psi}$ is A_2^3, then $\hat{f}_\eta^6$ is b^6.
7. If ψ is A_1^3 and $\dot{\psi}$ is A_2^1, then $\hat{f}_\eta^7$ is b^7.
8. If ψ is A_1^3 and $\dot{\psi}$ is A_2^2, then $\hat{f}_\eta^8$ is b^8.
9. If ψ is A_1^3 and $\dot{\psi}$ is A_2^3, then $\hat{f}_\eta^9$ is b^9.

where b^i, $i = 1, \ldots, 9$ are adjusted via RLS. The crisp output, which is an approximation of f_η, is

$$\hat{f}_\eta = \sum_{i=1}^{9} \xi_i b^i$$

where ξ_i, $i = 1, \ldots, 9$ are the same fuzzy basis functions as in (10.30) because the rule premises are identical.

The resulting estimated plant in companion form is

$$\ddot{y} \approx \hat{f}_\delta(\psi, \dot{\psi}) + \hat{f}_\eta(\psi, \dot{\psi})u = \sum_{i=1}^{9} \xi_i a^i + \sum_{i=1}^{9} \xi_i b^i u = \phi^T \theta \tag{10.31}$$

where

$$\phi^T = [\xi_1 \quad \xi_2 \quad \cdots \quad \xi_9 \quad \xi_1 u \quad \xi_2 u \quad \cdots \quad \xi_9 u]$$
$$\theta = \left[a^1 \quad a^2 \quad \cdots \quad a^9 \quad b^1 \quad b^2 \quad \cdots \quad b^9\right]^T$$

Since the estimated plant (10.31) is linear in the parameters θ, the unknown parameters a^i, b^i can be estimated via RLS (8.7) with the correction term equal to $[\ddot{\psi}(k) - \phi^T(k)\theta(k-1)]$.

If we desire the rod angle to track the output of a linear system with transfer function

$$\frac{Y_m(s)}{R(s)} = \frac{s+64}{s^2+16s+64} \tag{10.32}$$

excited by the reference signal

$$r(t) = 0.1\sin(3\pi t) \tag{10.33}$$

we apply the input

$$u(t) = \frac{1}{\hat{f}_\eta(t)}\left(-\hat{f}_\delta(t) + v(t)\right)$$

with

$$v(t) = -16\dot{\psi}(t) - 64\psi(t) + \dot{r}(t) + 64r(t)$$

The resulting response is shown in Figure 10.14. In this figure, we see that the rod angle tracks the reference model well.

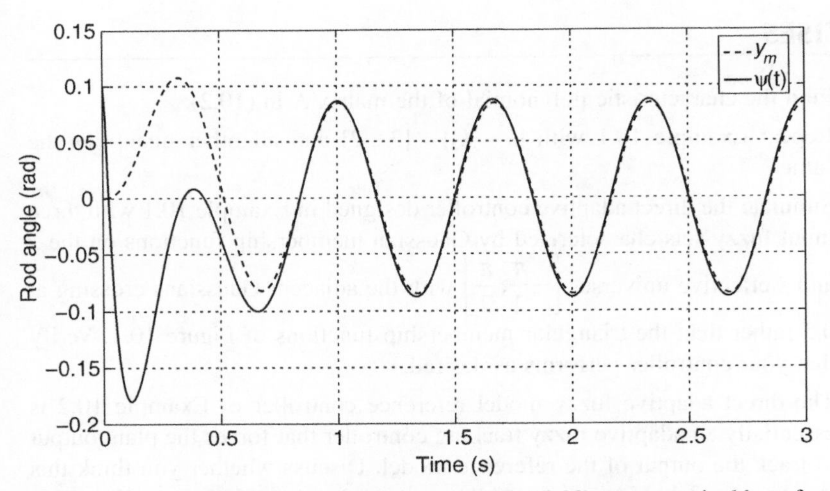

Figure 10.14. Rod angle $\psi(t)$ compared to model (10.32) output excited by reference signal (10.33) under indirect adaptive feedback linearization control, RLS parameter adaptation.

10.6 SUMMARY

This chapter shows how adaptive fuzzy controllers can be constructed. With the methods discussed here it is possible to form tracking and model reference controllers for unknown nonlinear systems.

Adaptive controllers are divided into two categories: **direct** and **indirect**. The **indirect adaptive control** strategy involves identifying the plant and simultaneously constructing a control law based on this identification. Thus the current estimate of the plant model is used in the calculation of the control law. This is called the **certainty equivalence principle**: the current plant model, even though incorrect, is used in the calculation of the controller. The strategy is that over time the erroneous controller converges to the correct controller as the identified plant approximation converges to the correct plant model. The direct control approach directly calculates the controller parameters without taking the extra step of identifying the plant.

For indirect adaptive control strategies in this book, we use either recursive least squares or gradient parameter estimation, because these two methods are recursive. We have discussed indirect adaptive **tracking** and **model reference** controllers with the identified model in the form of a T–S fuzzy system with input–output difference equation consequents, although other forms of consequents are also possible. We also discussed indirect adaptive feedback linearization in which the fuzzy model is in Mamdani form. We have also given an example of direct adaptive fuzzy tracking control.

It should be kept in mind that there are no formal proofs of stability for any of the methods presented in this chapter. Nevertheless, they have been used with success in many applications.

EXERCISES

10.1 Find the characteristic polynomial of the matrix Λ in (10.2).

10.2 Repeat Example 10.1 with $[k_1 \quad k_2] = [3 \quad 2]$ and all other quantities the same.

10.3 Simulate the direct adaptive controller designed in Example 10.1 with three input fuzzy sets characterized by Gaussian membership functions on the x and $\dot{x}$ effective universes $\left[-\frac{\pi}{2}, \frac{\pi}{2}\right]$ with the adjacent Gaussians crossing at 0.5 rather than the triangular membership functions of Figure 10.1. Verify that your controller performs as desired.

10.4 The direct adaptive fuzzy model reference controller of Example 10.2 is essentially an adaptive fuzzy tracking controller that forces the plant output to track the output of the reference model. Discuss whether you think this is a true model reference controller, or merely a tracking controller. Give reasons for your conclusions.

10.5 (a) Is the reference signal $r(t)$ of (10.16) periodic? Prove your answer.

(b) Is the gantry a minimum phase system? Explain.

10.6 In Example 10.2, assume that the model rules are of the form

R_{ip}: If $y(k)$ is A_1^i and $y(k-1)$ is A_2^i, then

$$\hat{y}^i(k+1) = \hat{a}_1^i y(k) + \hat{a}_2^i y(k-1) + \hat{b}_1^i u(k)$$

that is, $u(k)$ is present in the consequent, but $u(k-1)$ is not.

(a) How many unknown parameters are there?

(b) Write the recursive least-squares parameter estimate laws to estimate the unknown parameters.

(c) Write the rule in the controller fuzzy system that corresponds to Rule *ip* above.

10.7 Show that the control laws 10.26 and 10.27 result in a linear closed-loop system with transfer function given by 10.28.

10.8 Design an indirect adaptive fuzzy feedback linearization controller for the motor-driven robotic link of Example 10.5. Let the closed-loop system be linear with the transfer function

$$\frac{Y_m(s)}{R(s)} = \frac{10}{s^2+4s+10}$$

10.9 Repeat Example 10.5 with three input fuzzy sets characterized by Gaussian membership functions on the $y(k)$ and $y(k-1)$ effective universes $\left[-\frac{\pi}{2}, \frac{\pi}{2}\right]$ with adjacent Gaussians crossing at 0.5 rather than the triangular membership functions of Figure 10.9. Simulate and verify that your controller performs as desired.

10.10 Repeat Example 10.6 using gradient parameter estimation rather than recursive least squares. Simulate and verify that your controller performs as desired. (*Hint:* You will probably have to use Gaussian input membership functions rather than the triangular membership functions of Figs. 10.12 and 10.13).

REFERENCES

1. Friedland, B., *Advanced Control System Design*, Prentice-Hall, Englewood Cliffs, NJ, 1996.
2. Dorf, R. C. and Bishop, R. H., *Modern Control Systems* (10th ed.), Prentice-Hall, Englewood Cliffs, NJ, 2005.
3. Franklin, G. F., Powell, J. D., Emami-Naeini, *Feedback Control of Dynamic Systems* (5th ed.), Prentice Hall, Englewood Cliffs, NJ, 2006.
4. Ljung, L., *System Identification: Theory for the User*, Prentice-Hall, Englewood Cliffs NJ, 1987.
5. Yager, R. R., Ovchinnikov, S., Tong, R. M. and Nguyen, H. T., *Fuzzy Sets and Applications: Selected Papers* by L. A. Zadeh, John Wiley & Sons, Inc., New York, 1987.
6. Zadeh, L. A., "Fuzzy Sets, Information and Control," 8 (1965).
7. Mendel, J. M., *Uncertain Rule-Based Fuzzy Logic Systems: Introduction and New Directions*, Prentice-Hall, Upper Saddle River, NJ, 2001.
8. Klir, G. J. and Yuan, B., *Fuzzy Sets and Fuzzy Logic: Theory and Applications*, Prentice-Hall, Englewood Cliffs, NJ, 1995.
9. Passino, K. M. and Yurkovich, S., *Fuzzy Control*, Addison Wesley, 1998 (out of print). Available at http://www.eleceng.ohio-state.edu/~passino/FCbook.pdf
10. Ross, Timothy J., *Fuzzy Logic with Engineering Applications*, McGraw-Hill, New York, 1995.
11. Wang, L-X., *A Course in Fuzzy Systems and Control*, Prentice-Hall, Englewood Cliffs, NJ, 1997.
12. Wang, L-X., *Adaptive Fuzzy Systems and Control: Design and Stability Analysis*, Prentice-Hall, Englewood Cliffs, NJ, 1994.
13. Ying, Hao, *Fuzzy Control and Modeling: Analytical Foundations and Applications*, IEEE Press, Piscataway, NJ, 2000.
14. Branson, J. S. and Lilly, J. H., "Incorporation, Characterization, and Conversion of Negative Rules into Fuzzy Inference Systems," *IEEE Trans. Fuzzy Systems*, 9(2) (April 2001).
15. Lilly, J. H., "Evolution of a Negative-rule Fuzzy Obstacle Avoidance Controller for an Autonomous Vehicle," *IEEE Trans. Fuzzy Systems*, 15(4) (Aug. 2007).
16. Chan, S. W., Lilly, J. H., Repperger, D. W. and Berlin, J. E. , "Fuzzy PD+I Learning Control for a Pneumatic Muscle," Proc. 2003 IEEE Int. Conf. on Fuzzy Systems (FUZZ-IEEE'03), St. Louis, MO, May 2003.
17. Farinwata, S. S., Filev, D., Langari, R., Eds., *Fuzzy Control: Synthesis and Analysis*, John Wiley&Sons, Inc., Chichester, England, 1999.

Fuzzy Control and Identification, By John H. Lilly

18. Mamdani, E., *Adv. Linguistic Synth. Fuzzy Controllers, Int. J. Man-Machine Studies*, 8(6) (1976).
19. Close, C. M. and Frederick, D. K., *Modeling and Analysis of Dynamic Systems*, Houghton Mifflin, Boston, 1978.
20. Chang, X. and Lilly, J. H., "Evolutionary Design of a Fuzzy Classifier from Data," *IEEE Trans. Systems, Man, Cybernetics (Part B)*, 34(4) (Aug. 2004).
21. Vidyasagar, M., *Nonlinear Systems Analysis* (2nd ed.), Prentice-Hall, Englewood Cliffs, NJ, 1993.
22. Khalil, H. K., *Nonlinear Systems* (2nd ed.), Prentice-Hall, Englewood Cliffs, NJ, 1996.
23. Brown, M. and Harris, C., *Neurouzzy Adaptive Modelling and Control*, Prentice-Hall, New York, 1994.
24. de Silva, C. W., *Intelligent Control: Fuzzy Logic Applications*, CRC Press, Boca Raton, FL, 1995.
25. Takagi, T. and Sugeno, M., "Fuzzy Identification of Systems and its Applications to Modeling and Control," *IEEE Trans. Systems, Man, Cybernet.*, 15(1) (Jan. 1985).
26. Levine, William S., Ed., *The Control Handbook*, CRC Press, Boca Raton, FL, 1996.
27. Goodwin, G. C. and Sin, K. S., *Adaptive Filtering, Prediction, and Control*, Prentice-Hall, Englewood Cliffs, NJ, 1984.
28. Wang, L-X. and Mendel, J. M., "Fuzzy Basis Functions, Universal Approximation, and Orthogonal Least-squares Learning," *IEEE Trans. Neural Networks*, 3(5) (Sept. 1992).
29. Tanaka, K. and Wang, H. O., *Fuzzy Control Systems Design and Analysis: A Linear Matrix Inequality Approach*, John Wiley&Sons, Inc., New York, 2001.
30. Aracil, J. and Gordillo, F., *Stability Issues in Fuzzy Control*, Physica-Verlag, Heidelberg, 2000.
31. Boyd, S., El Ghaoui, L., Feron, E., and Balakrishnan, V., *Linear Matrix Inequalities in System and Control Theory*, SIAM, Philadelphia, 1994.
32. Sorenson, Harold W., *Parameter Estimation: Principles and Problems*, Marcel Dekker, New York, 1980.
33. Johnson, Jr., C. R., *Lectures in Adaptive Parameter Estimation*, Prentice-Hall, Englewood Cliffs, NJ, 1988.
34. Luenberger, D. G., *Linear and Nonlinear Programming*, Addison-Wesley, Reading, MA, 1984.
35. Chang, X. and Lilly, J. H., "Fuzzy Control for Pneumatic Muscle Tracking via Evolutionary Tuning," *Intell Automation Soft Comput*, 9(4) (Sept. 2003).
36. Astrom, K. J. and Wittenmark, B., *Computer Controlled Systems: Theory and Design* (3rd ed.), Prentice-Hall, Upper Saddle River, NJ, 1997.
37. Astrom, K. J. and Wittenmark, B., *Adaptive Control* (2nd ed.), Addison-Wesley, Reading, MA, 1995.
38. Ioannou, P. A. and Sun, J., *Robust Adaptive Control*, Prentice-Hall, Upper Saddle River, NJ, 1996.
39. Wellstead, P. E. and Zarrop, M. B., *Self Tuning Systems*, John Wiley & Sons, Inc., Chichester, England, 1991.
40. Layne, J. R. and Passino, K. M., "Fuzzy Model Reference Learning Control," *J. Intel. Fuzzy Systems*, 4(1) (1996).

APPENDIX

COMPUTER PROGRAMS

```
%tmfl evaluates left triangular membership function
%with corners m2 and m3 at point e
function tmfl=tmfl(e,m2,m3)
if e<=m2
   tmfl=1;

   elseif e>m2 & e<m3
      tmfl=-(1/(m3-m2))*e+m3/(m3-m2);

      else
         tmfl=0;

end
```

```
%tmf evaluates triangular membership function
%with corners m1, m2, and m3 at point e
function tmf=tmf(e,m1,m2,m3)
if e>m1 & e<=m2
   tmf=(1/(m2-m1))*e-m1/(m2-m1);

   elseif e>m2 & e<m3
      tmf=-(1/(m3-m2))*e+m3/(m3-m2);

      else
         tmf=0;

end
```

```
%tmfr evaluates right triangular membership function
%with corners m1 and m2 at point e
function tmfr=tmfr(e,m1,m2)
if e>m1 & e<=m2
   tmfr=(1/(m2-m1))*e-m1/(m2-m1);

   elseif e>m2
      tmfr=1;
```

```
    else
        tmfr=0;

end
```

```
%gmfl evaluates left gaussian membership function
%with center mid and spread sigma at point e
function gmfl=gmfl(e,mid,sigma)
if e<=mid
   gmfl=1;

   else
       gmfl=exp(-.5*((e-mid)/sigma)^2);

end
```

```
%gmf evaluates gaussian membership function
%with center mid and spread sigma at point e
function gmf=gmf(e,mid,sigma)
   gmf=exp(-.5*((e-mid)/sigma)^2);
```

```
%gmfr evaluates right gaussian membership function
%with center mid and spread sigma at point e
function gmfr=gmfr(e, mid, sigma)
if e>mid
   gmfr=1;

   else
       gmfr=exp(-.5*((e-mid)/sigma)^2);

end
```

```
%script fig2p1
%generates Figure 2.1
%single triangular mf

close all
clear all

T=-5:.01:40;    %range of temperatures for plotting

for i=1:length(T)
    mfT(i)=tmf(T(i),15,25,35);
end
plot(T,mfT,'k','linewidth',2),grid
axis([-5 40 0 1.2]);
```

```
%script fig2p2
%generates Figure 2.2
%single gaussian mf

close all
clear all

T=-5:.01:40;    %range of temperatures for plotting

for i=1:length(T)
   mfT(i)=gmf(T(i),25,5);
end

plot(T,mfT,'k','linewidth',2),grid
axis([-5 40 0 1.2]);
```

```
%script fig2p3
%generates figure 2.3
%illogical mf

T=-5:.1:40;    %range of temperatures for plotting

for i=1:length(T)

   if T(i)<20
      mu(i)=exp(-.5*((T(i)-20)/5)^2);

      elseif t(i)>=30
         mu(i)=exp(-.5*((T(i)-30)/5)^2);

         elseif t(i)>=20 & T(i)<30
           mu(i)=.5*(1.5+.5*cos(.1*2*pi*(T(i)-20)));

   end

end

plot(T,mu,'k','linewidth',2),grid
axis([-5 40 0 1.2])
```

```
%script fig2p4
%generates figure 2.4
%asymmetrical gaussian mf

close all
clear all

T=-5:.1:40;    %range of temperatures for plotting

for i=1:length(T)

    if T(i)<25
       mu(i)=gmf(T(i),25,7);
```

```
        elseif T(i)>=25
          mu(i)=gmf(T(i),25,3);
    end
end
plot(T,mu,‘k’,‘linewidth’,2),grid
axis([-5 40 0 1.2])
```

```
%script fig2p5
%generates Figure 2.5
%4 gaussian mfs
close all
clear all
T=linspace(-10,45,500); %range of temperatures for plotting
center=[5 15 25 35];    %centers of memberships
%spread for adjacent mfs crossing at .6
sigma=.5*(center(2)-center(1))/sqrt(-2*log(.6));
for i=1:length(T)
    gmf1(i)=gmfl(T(i),center(1),sigma);
    gmf2(i)=gmf(T(i),center(2),sigma);
    gmf3(i)=gmf(T(i),center(3),sigma);
    gmf4(i)=gmfr(T(i),center(4),sigma);
end
plot(T,gmf1,‘k’,T,gmf2,‘k’,T,gmf3,‘k’,T,gmf4,‘k’,…‘linewidth’,
      2),grid;
axis([-10 45 0 1.2])
```

```
%fig2p12
%generates figure 2.12
%Cartesian product of 2 fuzzy sets
close all
clear all
S=0:1:50;   %range of wind speeds for plotting
T=-10:1:40; %range of temperatures for plotting
for i=1:length(T)
    for j=1:length(S)
        mu(i,j)=gmf(T(i),25,5)*gmf(S(j),15,4);
    end
end
colormap(white)
surface(T,S,mu’),view(43,15),grid
```

```
%script fig3p2
%generates Figure 3.2
%4 triangular mfs

close all
clear all

center=[0 10 20 30]; %membership centers

Temp=-5:.1:35;    %range of temperatures for plotting

for i=1:length(Temp)
    COLD(i)=tmfl(Temp(i),center(1),center(2));
    COOL(i)=tmf(Temp(i),center(1),center(2),center(3));
    WARM(i)=tmf(Temp(i),center(2),center(3),center(4));
    HOT(i)=tmfr(Temp(i),center(3),center(4));
end

plot(Temp,COLD,'k',Temp,COOL,'k',Temp,WARM,'k',...
     Temp,HOT,'k','linewidth',2),grid
axis([-5 35 0 1.2])
```

```
%script fig3p6
%generates Figure 3.6 using min T-norm and COG
%defuzzification
%windchill input-output characteristic

close all;    %close all open plots
clear all;    %clear all variables in the work space

numpts=50;                  %number of values of temp and
                            %windspeed at which to evaluate mfs

midT=[0 10 20 30];          %mid points of mfs on temp ud
midS=[5 15 25];             %mid points of mfs on windspeed ud
midW=[-25 -10 5 20 35];     %mid points of mfs on windchill ud

w=midW(3)-midW(1);    %widths of bases of windchill triangles

%centers of area of output mfs
FRIG=midW(1);
COLD=midW(2);
COOL=midW(3);
COMF=midW(4);
WARM=midW(5);

%rule base
rule=[COOL COLD FRIG;...
      COMF COOL COLD;...
      WARM COMF COOL;...
      WARM WARM COMF];
```

```
%define numpts points equally spaced T and S uds
T=linspace(-5,35,numpts);
S=linspace(0,30,numpts);

%for each of the numpts^2 points defined in temp x windspeed
%plane, calculate the corresponding wind chill
for i=1:numpts
for j=1:numpts

%calculate degree of membership in the 4 temp fuzzy sets
        mfT=[tmfl(T(i),midT(1),midT(2))…
             tmf(T(i),midT(1),midT(2),midT(3))…
             tmf(T(i),midT(2),midT(3),midT(4))
             tmfr(T(i),midT(3),midT(4))];

%calculate degree of membership in the 3 windspeed fuzzy
%sets
         mfS=[tmfl(S(j),midS(1),17.5)…
             tmf(S(j),2.5,midS(2),27.5)…
             tmfr(S(j),12.5,midS(3))];

%calculate the degree of firing of each of the 12 rules
%using min T-norm.   prem is a 4x3 matrix
        for k=1:4
          for l=1:3
             prem(k,l)=min(mfT(k),mfS(l));
          end
        end

%the heights of the implied fuzzy sets (trapezoids)
%h is a 4x3 matrix
        h=prem;

%the areas of the implied fuzzy sets (trapezoids)
%h is a 4x3 matrix
        areaimp=w*(h-h.^2/2);

        num=sum(sum(areaimp.*rule));
        den=sum(sum(areaimp));

%the crisp output (windchill) corresponding to the
%ith value of temperature and the jth value of windspeed

        windchill(i,j)=num/den;

   end
end
```

```
axis([-5 35 0 30 -30 40]);
colormap(white)
surface(T,S,windchill'),view(200,15),grid
xlabel('Temperature')
ylabel('Wind speed')
zlabel('Windchill')
```

```
%script fig4p12
%generates Figure 4.12, product T-norm, CA defuzzification
%states: x(1)=x, x(2)=xdot
%4th order Runge-Kutta integration
%response of ball and beam under fuzzy control

close all
clear all

%g0=1;    %scaling gain
%g1=1;    %scaling gain
%h=1;     %scaling gain

%g0=1;    %scaling gain
%g1=18;   %scaling gain
%h=1;     %scaling gain

g0=3;     %scaling gain
g1=18;    %scaling gain
h=1;      %scaling gain

%g0=3;    %scaling gain
%g1=18;   %scaling gain
%h=7;     %scaling gain

X1=[];    %placeholder for X1 array
V=[];     %placeholder for V array
T=[];     %placeholder for T array

x=[-.4 0];   %initial conditions
dx=[0 0];

t=0;            %initial time
dt=.001;        %time step
tfinal=10-dt;   %final time

k=.01;          %motor shaft angle-to-voltage ratio
g=9.81;         %acceleration of gravity (m/s^2)

%triangle midpoints on e, c, and F universes
mid=[-1 -.5 0 .5 1];
mide=.5*mid;    %triangle mid-points on e universe
midc=4*mid;     %triangle mid-points on c universe
midv=10*mid;    %triangle mid-points on v universe
base=midv(3)-midv(1); %base of output triangles
```

```
%locations of output singletons
NL=midv(1);
NS=midv(2);
ZE=midv(3);
PS=midv(4);
PL=midv(5);

%rule base
rule=[NL NL NL NS ZE; …
     NL NL NS ZE PS; …
     NL NS ZE PS PL; …
     NS ZE PS PL PL; …
     ZE PS PL PL PL];

while t<=tfinal

   x1previous=x(1);    %needed to approximate dx/dt

   for n=1:4   %needed for RK4 algorithm

      %inputs to fuzzy controller
      e=-x(1);
      c=-(x(1)-x1previous)/dt;    %approximation of dx/dt

      %membership function values corresponding to e above
      mfe=[tmfl(g0*e,mide(1),mide(2))…
           tmf(g0*e,mide(1),mide(2),mide(3))…
           tmf(g0*e,mide(2),mide(3),mide(4))…
           tmf(g0*e,mide(3),mide(4),mide(5))…
           tmfr(g0*e,mide(4),mide(5))];

      %membership function values corresponding to c above
      mfc=[tmfl(g1*c,midc(1),midc(2))…
           tmf(g1*c,midc(1),midc(2),midc(3))…
           tmf(g1*c,midc(2),midc(3),midc(4))…
           tmf(g1*c,midc(3),midc(4),midc(5))…
           tmfr(g1*c,midc(4),midc(5))];

      %calculate degrees of firing of all rules
      prem=mfe'*mfc;

      %calculate crisp output via CA defuzzification
      num=sum(sum(rule.*prem));
      den=sum(sum(prem));
      volts=h*num/den;    %crisp output

%     ball and beam model
      dx(1)=x(2);
      dx(2)=g*sin(k*volts);
```

```
%RK4 generic code between dotted lines
%-------------------------------------------------------------------
      if n==1
         z=dx;
         v=x;
         x=v+0.5*dt*dx;
         t=t+0.5*dt;

         elseif n==2
             z=z+2*dx;
             x=v+0.5*dt*dx;

             elseif n==3
                 z=z+2*dx;
                 x=v+dt*dx;
                 t=t+0.5*dt;

                 elseif n==4
                     x=v+(dt/6)*(z+dx);

      end
%-------------------------------------------------------------------

   end

   X1=[X1 x(1)];  %save current state x(1) for plotting
   V=[V volts];   %save current input (V) for plotting
   T=[T t];

end

plot(T,X1,'k','linewidth',2),grid
```

```
%script fig4p18
%generates Figure 4.18
%Gaussian input memberships, product T-norm
%input-output characteristic of fuzzy controller for
%ball and beam

close all;    %close all open plots
clear all;    %clear all variables in the work space

numpts=40;    %number of values of e and edot at which to
              %evaluate mfs

mids=[-1 -.5 0 .5 1]; %gaussian midpoints on error universe
mide=.5*mids;
midc=4*mids;
midf=10*mids;

%spreads for adjacent mfs crossing at .5
sigmae=.5*(mide(2)-mide(1))/sqrt(-2*log(.5));
sigmac=.5*(midc(2)-midc(1))/sqrt(-2*log(.5));
```

```
%linguistic values for output mfs
NL=midf(1);
NS=midf(2);
ZE=midf(3);
PS=midf(4);
PL=midf(5);

%rule base
rule=[NL NL NL NS ZE; ...
     NL NL NS ZE PS; ...
     NL NS ZE PS PL; ...
     NS ZE PS PL PL; ...
     ZE PS PL PL PL];

%define numpts points equally spaced e and c uds

e=linspace(-.55,.55,numpts);
c=linspace(-4.4,4.4,numpts);

for i=1:numpts
    for j=1:numpts

%membership function values corresponding to e above
        mfe=[gmfl(e(i),mide(1),sigmae) ...
             gmf(e(i),mide(2),sigmae) ...
             gmf(e(i),mide(3),sigmae) ...
             gmf(e(i),mide(4),sigmae) ...
             gmfr(e(i),mide(5),sigmae)];

%membership function values corresponding to c above
        mfc=[gmfl(c(j),midc(1),sigmac) ...
             gmf(c(j),midc(2),sigmac) ...
             gmf(c(j),midc(3),sigmac) ...
             gmf(c(j),midc(4),sigmac) ...
             gmfr(c(j),midc(5),sigmac)];

             prem=mfe'*mfc;   %degrees of firing of all rules

%calculate crisp output
             num=sum(sum(rule.*prem));
             den=sum(sum(prem));
             ufuzzy(i,j)=num/den;

    end
end

axis([-0.55 0.55 -4.4 4.4 -12 12]);
colormap(white)
surface(e,c,ufuzzy'),view(20,15),grid
```

```
%script fig4p26
%generates Figure 4.26
%input-output characteristics of
%PD and fuzzy controllers for inverted pendulum

close all;    %close all open plots
clear all;    %clear all variables in the work space

numpts=20;    %number of values of e and edot at which to
              %evaluate mfs

P=30; %proportional gain
D=5; %derivative gain

maxf=3; %max control effort resulting from pd with above P
        %and D gains
mids5=[-1 -.5 0 .5 1];
mids9=[-2 -1.5 mids5 1.5 2];
mide=(2*maxf/P)*mids5;    %triangle mid-points on e universe
midc=(2*maxf/D)*mids5;    %triangle mid-points on c universe
midf=2*maxf*mids9;    %unredesigned singleton locations on
                      %F universe

%redesigned singleton locations on F universe
NLLL=2.5*midf(1);
NLL=2*midf(2);
NL=1.5*midf(3);
NS=midf(4);
ZE=midf(5);
PS=midf(6);
PL=1.5*midf(7);
PLL=2*midf(8);
PLLL=2.5*midf(9);

%generic rule base for equivalent fuzzy PD controller
rule=[NLLL NLL NL   NS    ZE; ...
      NLL  NL  NS   ZE    PS; ...
      NL   NS  ZE   PS    PL; ...
      NS   ZE  PS   PL    PLL; ...
      ZE   PS  PL   PLL   PLLL];

%define numpts equally spaced points on e and c universes
e=linspace(-.2,.2,numpts);
c=linspace(-1.2,1.2,numpts);

for i=1:numpts
    for j=1:numpts
```

```
        %membership function values corresponding to e above
        mfe=[tmfl(e(i),mide(1),mide(2)) ...
              tmf(e(i),mide(1),mide(2),mide(3)) ...
              tmf(e(i),mide(2),mide(3),mide(4)) ...
              tmf(e(i),mide(3),mide(4),mide(5)) ...
              tmfr(e(i),mide(4),mide(5))];

        %membership function values corresponding to c above
        mfc=[tmfl(c(j),midc(1),midc(2)) ...
              tmf(c(j),midc(1),midc(2),midc(3)) ...
              tmf(c(j),midc(2),midc(3),midc(4)) ...
              tmf(c(j),midc(3),midc(4),midc(5)) ...
              tmfr(c(j),midc(4),midc(5))];

        prem=mfe'*mfc; %degree of firing of all rules

%crisp output of nonlinear fuzzy controller
        num=sum(sum(rule.*prem));
        den=sum(sum(prem));
        ufuzzy(i,j)=-num/den;

%output of nonfuzzy PD controller
        upid(i,j)=-(P*e(i)+D*c(j));

    end
end

hold on
axis(1.1*[-0.2 0.2 -1.2 1.2 -30 30]);
colormap(white)
surface(e,c,ufuzzy')
surface(e,c,upid'),view(200,10),grid
```

```
%script Ex5p1
%Example 5.1 - Plots output of forced pendulum, nonlinear
%differential equation model

close all
clear all

M=1;         %pendulum mass (kg)
l=1;         %1/2 of pendulum length (m)
I=M*l^2;     %moment of inertia of pendulum
B=1;         %coefficient of friction at attach point
g=9.81;      %acceleration of gravity (m/sec^2)

X1=[];       %placeholder for x(1) array
X2=[];
U=[];
T=[];
```

```
t=0;          %initial time
dt=.001;      %step size for RK4 routine

x=[0;0];      %initial conditions
dx=[0;0];

final time=10-dt;

while t<finaltime

    for n=1:4 %part of generic RK4 code

        %specification of input
        if t<=2
            u=2*sin(2*pi*t);

            else
                u=0;

        end

        %mathematical model of forced pendulum
        dx(1)=x(2);
        dx(2)=(-B*x(2)-M*g*l*sin(x(1))+u)/I;

%generic RK4 code between dotted lines
%-----------------------------------------------------------------
        if n==1
            z=dx;
            v=x;
            x=v+0.5*dt*dx;
            t=t+0.5*dt;

            elseif n==2
                z=z+2*dx;
                x=v+0.5*dt*dx;

                elseif n==3
                    z=z+2*dx;
                    x=v+dt*dx;
                    t=t+0.5*dt;

                    elseif n==4
                        x=v+(dt/6)*(z+dx);

        end
%-----------------------------------------------------------------

    end

    X1=[X1 x(1)];     %save for plotting
    X2=[X2 x(2)];
    U=[U u];
    T=[T t];
```

```
end
plot(T,X1,'k','linewidth',2),grid
```

```
%script Ex5p3
%Example 5.3 - Plots output of forced pendulum,
%discrete-time state equation model

t=0:.01:10;     %sample times
x1=0;           %initial condition
x2=0;           %initial condition
X1=[];          %used for plotting

for k=1:length(t)

   %input torque to pendulum
   if t(k)<=2
      u=2*sin(2*pi*t(k));
      else
         u=0;

   end

   %discrete state equation model of pendulum
   x1n=x1+.01*x2;
   x2n=-.098*x1+.99*x2+.01*u;

   %update states for next iteration
   x1=x1n;
   x2=x2n;

   %save x1 values for plotting
   X1=[X1 x1];

end

plot(t,X1(1:length(t)),'k','linewidth',3),grid
```

```
%script Ex5p4
%Example 5.4 - Plots output of forced pendulum, linear
%input-output %difference equation model

dt=.01;
t=0:dt:10;      %sample times
ykm1=0;         %initial value of y(k-1)
ykm2=0;         %initial value of y(k-2)

Y=[];           %placeholder for y(k) array

for k=1:length(t)

    if t(k)>=0+2*dt & t(k)<=2+2*dt
       ukm2=2*sin(2*pi*t(k-2));
```

```
        else
            ukm2=0;

    end

    %input-output difference equation
    yk=1.99*ykm1-.991*ykm2+.0001*ukm2;

    ykm2=ykm1; %when time advances, y(k-1) becomes y(k-2)
    ykm1=yk;   %when time advances, y(k) becomes y(k-1)

    Y=[Y yk];  %save values of y(k) for plotting

end

plot(t,Y,'k','linewidth',3),grid
```

```
%script fig6p6
%creates Figure 6.6
%plots the states of a T-S fuzzy system that interpolates
%four time-invariant continuous-time linear systems

close all
clear all

%corners of triangle mfs on x1 and x2 uds
midx1=[-1 1];
midx2=[-2 2];

%coefficients in consequents of TS fuzzy rules
A1=[0 1;-2 -2];
A2=[0 1;-1 -2];
A3=[0 1;-3 -2];
A4=[1 1;-2 -2];
b1=[0;2];
b2=[2;2];
b3=[-2;2];
b4=[0;-2];

x=[0;0];    %initial conditions of state
dx=[0;0];   %initial conditions of derivatives of states

%placeholders for plotting saved variables
X1=[];
X2=[];
T=[];

t=0;       %initial time
dt=.01;    %step size of RK4 routine

finaltime=10-dt;

while t<finaltime
```

```
    for n=1:4    %part of RK4 routine

        u=3*sin(pi*t);    %input

%individual mf values corresponding to current x1 and x2
        mux1=[tmfl(x(1),midx1(1),midx1(2)) ...
              tmfr(x(1),midx1(1),midx1(2))];
        mux2=[tmfl(x(2),midx2(1),midx2(2)) ...
              tmfr(x(2),midx2(1),midx2(2))];

        %degrees of firing of all rules, product T-norm
        mu=mux1'*mux2;

        %fuzzy basis functions
        zeta=mu/sum(sum(mu));
        %overall A and b corresponding to current x1, x2

        A=zeta(1,1)*A1+zeta(1,2)*A2+...
          zeta(2,1)*A3+zeta(2,2)*A4;
        b=(zeta(1,1)*b1+zeta(1,2)*b2+...
           zeta(2,1)*b3+zeta(2,2)*b4);

        %overall DE
        dx=A*x+b*u;

%RK4 algorithm between dotted lines-----------------

        if n==1
            z=dx;
            v=x;
            x=v+0.5*dt*dx;
            t=t+0.5*dt;

            elseif n==2
                z=z+2*dx;
                x=v+0.5*dt*dx;

                elseif n==3
                    z=z+2*dx;
                    x=v+dt*dx;
                    t=t+0.5*dt;

                    elseif n==4
                        x=v+(dt/6)*(z+dx);

        end
%----------------------------------------------------------

    end
    X1=[X1 x(1)];
    X2=[X2 x(2)];
    T=[T t];
```

```
end

plot(T,X1,T,X2,'linewidth',2),grid
xlabel('Time (seconds)')
ylabel('States')
```

```
%script fig6p7
%creates Figure 6.7
%plots the states of a T-S fuzzy system that interpolates
%four time-invariant discrete-time linear systems

close all
clear all

dt=.2;   %sample time
finaltime=9;
t=0:dt:finaltime;   %times at which to calculate the states

%corners of triangle mfs on x1 and x2 uds
midx=[-1 1];

%coefficients in consequents of TS fuzzy rules
A1=[0.1 0;-0.2 0.2];
A2=[0 0.1;1 -0.2];
A3=[-0.3 1;0 0.2];
A4=[0.1 1;0.2 0.5];
b1=[0;1];
b2=[0.1;1];
b3=[-0.1;1];
b4=[0;-1];

x=[0;0];   %initial conditions of state
%placeholders for plotting saved variables
X1=[];
X2=[];
U=[];

for k=1:length(t)

   %individual mf values corresponding to current x1 and x2
   mux1=[tmfl(x(1),midx(1),midx(2)) ...
         tmfr(x(1),midx(1),midx(2))];
   mux2=[tmfl(x(2),midx(1),midx(2)) ...
         tmfr(x(2),midx(1),midx(2))];

   %degrees of firing of all rules, product T-norm
   mu=mux1'*mux2;

   %fuzzy basis functions
   zeta=mu/sum(sum(mu));
```

```
    %overall A and b corresponding to current x1, x2
    A=(zeta(1,1)*A1+zeta(1,2)*A2+zeta(2,1)*A3+zeta(2,2)*A4);
    b=(zeta(1,1)*b1+zeta(1,2)*b2+zeta(2,1)*b3+zeta(2,2)*b4);

    u=sin(.5*pi*t(k));    %new u(k)
    %overall difference equation
    x=A*x+b*u;

    X1=[X1 x(1)];
    X2=[X2 x(2)];
    U=[U u];

end

plot(t,X1,'o',t,X2,'o',t,U,'s'),grid
legend('x1(t)','x2(t)','u(t)');
xlabel('Time (seconds)')
ylabel('States')
axis([0 9 -1.2 1.2])
```

```
%script fig6p8
%creates Figure 6.8
%time-varying discrete system in input-output difference
%equation form from a four-rule TS fuzzy system

close all
clear all

dt=1;    %sample time
finaltime=60;

t=0:dt:finaltime;    %times at which to calculate the states

%corners of Gaussian mfs on y(k) and y(k-1) uds
midy=[-1 1];

%spread for adjacent Gaussians crossing at 0.5
sigma=.5*(midy(2)-midy(1))/sqrt(-2*log(.5));

ukm1=0; %initial u(k-1)
uk=0;   %initial u(k)
ykm1=0; %initial y(k-1)
yk=0;   %initial y(k)

%placeholders for y and u arrays
Y=[];
U=[];

%coefficients of y(k) in rule consequents
param1=[.5 .4 .2 .8];
```

```
%coefficients of y(k-1) in rule consequents
param2=[-.5 -.8 .5 -.6];

%coefficients of u(k) in rule consequents
param3=[1 1.2 1.5 -1.5];

%coefficients of u(k-1) in rule consequents
param4=[.6 -1 -.7 1];

for k=1:length(t)

   %calculate degrees of membership in all input fuzzy sets
   muyk=[gmfl(yk,midy(1),sigma) ...
        gmfr(yk,midy(2),sigma)];
   muykm1=[gmfl(ykm1,midy(1),sigma) ...
          gmfr(ykm1,midy(2),sigma)];

   %calculate degrees of firing of all rules
   mu=muyk'*muykm1;

   %calculate fuzzy basis functions
   zeta=[mu(1,:)/sum(sum(mu)) mu(2,:)/sum(sum(mu))];

   %calculate y(k+1)
   ykp1=(param1*zeta')*yk+(param2*zeta')*ykm1+...
        (param3*zeta')*uk+(param4*zeta')*ukm1;

   %update y(k-1), y(k), u(k-1), u(k)
   ykm1=yk;
   yk=ykp1;
   ukm1=uk;
   uk=sin(.05*pi*t(k));
   %save y(k) and u(k) for plotting
   Y=[Y yk];
   U=[U uk];

end

hold on
plot(t,U,'s','linewidth',2,'color',[.7.7.7])
plot(t,Y,'o','linewidth',2,'color','k'),grid
legend('\itu(t)','\ity(t)','Location','Southwest');
xlabel('Time (seconds)')
ylabel('States')
axis([0 60 -2 1.2])
```

```
%script ex7p1
%parallel distributed pole placement for a CT nonlinear
%system

close all
clear all
```

```
%corners of saturated Gaussian mfs on x1, x2 uds
mid=[-1 1];
cross=0.5; %crossing point of adjacent mfs

%spread for adjacent mfs crossing at 0.5
sigma=.5*(mid(2)-mid(1))/sqrt(-2*log(cross));

%coefficients in consequents of plant TS fuzzy rules
A1=[0 1;-2 2];
A2=[0 1;-3 2];
A3=[0 1;-2 1];
A4=[0 1;-3 1];
b1=[0;1];
b2=[0;1];
b3=[0;1];
b4=[0;1];

%coefficients in consequents of controller TS fuzzy
rules k1=[0 4];
k2=[-1 4];
k3=[0 3];
k4=[-1 3];

x=[1;-1];    %initial condition of state
dx=[0;0];    %initial condition of derivative of state
X1=[];
X2=[];
U=[];
T=[];

t=0;         %initial time
dt=0.01;     %step size of RK4 routine
finaltime=6-dt;

while t<finaltime

   for n=1:4   %part of RK4 routine

%degrees of belongingness in fuzzy sets on x1, x2 universes
      mu11=gmfl(x(1),mid(1),sigma);
      mu12=gmfr(x(1),mid(2),sigma);
      mu21=gmfl(x(2),mid(1),sigma);
      mu22=gmfr(x(2),mid(2),sigma);
%degrees of firing of rules
      mu1=mu11*mu21;
      mu2=mu11*mu22;
      mu3=mu12*mu21;
      mu4=mu12*mu22;
```

```
%fuzzy basis functions
      zeta1=mu1/(mu1+mu2+mu3+mu4);
      zeta2=mu2/(mu1+mu2+mu3+mu4);
      zeta3=mu3/(mu1+mu2+mu3+mu4);
      zeta4=mu4/(mu1+mu2+mu3+mu4);

%overall A, b, and k
      A=(zeta1*A1+zeta2*A2+zeta3*A3+zeta4*A4);
      b=(zeta1*b1+zeta2*b2+zeta3*b3+zeta4*b4);
      k=(zeta1*k1+zeta2*k2+zeta3*k3+zeta4*k4);

%feedback control law
      u=-k*x;

%closed-loop system
      dx=A*x+b*u;

%RK4 routine from here to dotted line
      if n==1
         z=dx;
         v=x;
         x=v+0.5*dt*dx;
         t=t+0.5*dt;

         elseif n==2
            z=z+2*dx;
            x=v+0.5*dt*dx;

            elseif n==3
               z=z+2*dx;
               x=v+dt*dx;
               t=t+0.5*dt;

               elseif n==4
                  x=v+(dt/6)*(z+dx);

      end
%-----------------------------------------------------------

   end

%save variables for plotting
   X1=[X1 x(1)];
   X2=[X2 x(2)];
   U=[U u];
   T=[T t];

end

plot(T,X1,'k',T,X2,'k','linewidth',3),grid
```

```
%script ex7p2
%parallel distributed pole placement for a DT nonlinear
%system

close all
clear all

%corners of saturated Gaussian mfs on x1, x2 uds
mid=[-1 1];

cross=0.5; %crossing point of adjacent mfs

%spread for adjacent mfs crossing at cross
sigma=.5*(mid(2)-mid(1))/sqrt(-2*log(cross));

%matrices in consequents of plant TS fuzzy rules
A1=[0 1;-1 1];
A2=[1 1;-1 2];
A3=[0 1;-.2 1];
A4=[1 1;-.3 1];
b1=[0;1];
b2=[0;1];
b3=[0;1];
b4=[0;1];

%feedback gains in consequents of controller TS fuzzy rules
k1=[-1.25 1];
k2=[-0.25 3];
k3=[-0.45 1];
k4=[0.45 2];

x=[1;-1];    %initial condition of state

%placeholders for plotting states
X1=[];
X2=[];
U=[];
T=[];

t=0;
dt=0.1;    %sample time

finaltime=1;

while t<finaltime

%degrees of belongingness in fuzzy sets on x1, x2 universes
       mu11=gmfl(x(1),mid(1),sigma);
       mu12=gmfr(x(1),mid(2),sigma);
       mu21=gmfl(x(2),mid(1),sigma);
       mu22=gmfr(x(2),mid(2),sigma);
```

```
%degrees of firing of rules
      mu1=mu11*mu21;
      mu2=mu11*mu22;
      mu3=mu12*mu21;
      mu4=mu12*mu22;

%fuzzy basis functions
      zeta1=mu1/(mu1+mu2+mu3+mu4);
      zeta2=mu2/(mu1+mu2+mu3+mu4);
      zeta3=mu3/(mu1+mu2+mu3+mu4);
      zeta4=mu4/(mu1+mu2+mu3+mu4);

%overall A, b, and k
      A=(zeta1*A1+zeta2*A2+zeta3*A3+zeta4*A4);
      b=(zeta1*b1+zeta2*b2+zeta3*b3+zeta4*b4);
      k=(zeta1*k1+zeta2*k2+zeta3*k3+zeta4*k4);

%feedback control law
      u=-k*x;

%closed-loop system

%save variables for plotting
      X1=[X1 x(1)];
      X2=[X2 x(2)];
      U=[U u];
      T=[T t];

      t=t+dt;

      x=A*x+b*u;   %nonlinear system model

end

hold on

plot(T,X1,'ko',T,X2,'ko','markersize',10,'linewidth',3),grid
legend('\itx1(t)','\itx2(t)')
axis([0 1 -1.2 1.2])
```

```
%script ex7p3
%Example 7.3, parallel distributed one-step-ahead tracking

close all
clear all

finaltime=20;
k=0;   %initial time

%corners of triangle mfs on y(k) and y(k-1) uds
midy=[-1 1];
```

```
ukm1=0;      %initial u(k-1)=u(-1)
uk=0;        %initial u(k)=u(0)
ykm1=-1;     %initial y(k-1)=y(-1)
yk=1;        %initial y(k)=y(0)

%placeholders for vectors saved for plotting
Y=[];
R=[];
K=[];

while k<=finaltime

   %evaluate memberships for present y(k), y(k-1)
   mu11=tmfl(yk,midy(1),midy(2));
   mu12=tmfr(yk,midy(1),midy(2));
   mu21=tmfl(ykm1,midy(1),midy(2));
   mu22=tmfr(ykm1,midy(1),midy(2));

   %calculate degrees of firing of all rules
   mu1=mu11*mu21;
   mu2=mu11*mu22;
   mu3=mu12*mu21;
   mu4=mu12*mu22;

   %calculate basis functions
   zeta1=mu1/(mu1+mu2+mu3+mu4);
   zeta2=mu2/(mu1+mu2+mu3+mu4);
   zeta3=mu3/(mu1+mu2+mu3+mu4);
   zeta4=mu4/(mu1+mu2+mu3+mu4);

   %calculate one-step-ahead reference signal to be tracked
   rk=0.5*sin(0.2*pi*k);
   rkp1=0.5*sin(0.2*pi*(k+1));

   %calculate consequents of controller fuzzy system
   ccons1=-0.6*ukm1-1.5*yk+0.4*ykm1+rkp1;
   ccons2=(-ukm1-0.4*yk+1.8*ykm1+rkp1)/1.2;
   ccons3=(-0.7*ukm1-1.2*yk-0.5*ykm1+rkp1)/1.5;
   ccons4=(-ukm1-0.8*yk+1.6*ykm1+rkp1)/1.5;

   %calculate control signal u(k)
   uk=zeta1*ccons1+zeta2*ccons2+zeta3*ccons3+zeta4*ccons4;

   %calculate consequents of plant fuzzy system
   pcons1=1.5*yk-0.4*ykm1+uk+0.6*ukm1;
   pcons2=0.4*yk-1.8*ykm1+1.2*uk+ukm1;
   pcons3=1.2*yk+0.5*ykm1+1.5*uk+0.7*ukm1;
   pcons4=0.8*yk-1.6*ykm1+1.5*uk+ukm1;

   %calculate y(k+1) of plant fuzzy system
```

```
    ykp1=zeta1*pcons1+zeta2*pcons2+...
         zeta3*pcons3+zeta4*pcons4;

    %save y(k) and r(k) for plotting
    Y=[Y yk];
    R=[R rk];
    K=[K k];

    %update u and y for next iteration
    ukm1=uk;
    ykm1=yk;
    yk=ykp1;

    k=k+1;

end

%plot tracking error
%plot(K,Y,'ko',K,R,'ksq'),gridplot(K,Y,'ko',K,R,'ksq',
'markersize',10,'linewidth',3),grid
%plot(K,Y-R,'ko','markersize',10,'linewidth',3),grid
legend('\ity(t)','\itr(t)')
axis([0 20 -0.6 1.2])
```

```
%script ex7p4
%Example 7.4, parallel distributed model reference control

close all
clear all

finaltime=20;
k=0;    %initial time

%corners of saturated triangle mfs on y(k) and y(k-1) uds
midy=[-1 1];

ukm1=0;     %initial plant input u(k-1)=u(-1)
uk=0;       %initial plant input u(k)=u(0)
ykm1=-1;    %initial plant output y(k-1)=y(-1)
yk=1;       %initial plant output y(k)=y(0)
Ykm1=0;     %initial reference model output Y(k-1)
Yk=0;       %initial reference model output Y(k)
rk=0;       %initial reference model input r(k)

%placeholders for vectors saved for plotting
Y=[];
YM=[];
K=[];

while k<=finaltime
```

```
%evaluate memberships for present y(k), y(k-1)
mu11=tmfl(yk,midy(1),midy(2));
mu12=tmfr(yk,midy(1),midy(2));
mu21=tmfl(ykm1,midy(1),midy(2));
mu22=tmfr(ykm1,midy(1),midy(2));

%calculate degrees of firing of all rules
mu1=mu11*mu21;
mu2=mu11*mu22;
mu3=mu12*mu21;
mu4=mu12*mu22;

%calculate basis functions
zeta1=mu1/(mu1+mu2+mu3+mu4);
zeta2=mu2/(mu1+mu2+mu3+mu4);
zeta3=mu3/(mu1+mu2+mu3+mu4);
zeta4=mu4/(mu1+mu2+mu3+mu4);

%calculate reference input
rkm1=rk;
rk=0.5*sin(0.2*pi*k);

%calculate reference model output
Ykp1=-0.5*Yk-0.5*Ykm1+2*rk+0.6*rkm1;

%calculate consequents of controller fuzzy system
ccons1=-0.6*ukm1-2*yk-0.1*ykm1+2*rk+0.6*rkm1;
ccons2=(-ukm1-0.9*yk+1.3*ykm1+2*rk+0.6*rkm1)/1.2;
ccons3=(-0.7*ukm1-1.7*yk-ykm1+2*rk+0.6*rkm1)/1.5;
ccons4=(-ukm1-1.3*yk+1.1*ykm1+2*rk+0.6*rkm1)/1.5;

%calculate control signal u(k)
uk=zeta1*ccons1+zeta2*ccons2+zeta3*ccons3+zeta4*ccons4;

%calculate consequents of plant fuzzy system
pcons1=1.5*yk-0.4*ykm1+uk+0.6*ukm1;
pcons2=0.4*yk-1.8*ykm1+1.2*uk+ukm1;
pcons3=1.2*yk+0.5*ykm1+1.5*uk+0.7*ukm1;
pcons4=0.8*yk-1.6*ykm1+1.5*uk+ukm1;

%calculate y(k+1) of plant fuzzy system
ykp1=zeta1*pcons1+zeta2*pcons2+...
     zeta3*pcons3+zeta4*pcons4;

%save yplant(k+1) and Ymodel(k+1) for plotting
Y=[Y yk];
YM=[YM Yk];
K=[K k];
```

```
    %update u, yplant, and Ymodel for next iteration
    ukm1=uk;
    ykm1=yk;
    yk=ykp1;
    Ykm1=Yk;
    Yk=Ykp1;

    k=k+1;

end

%plot tracking error
plot(K,Y,'ko',K,YM,'ksq','markersize',10,'linewidth',3),grid
%plot(K,Y-YM,'ko','markersize',10,'linewidth',3),grid
legend('\ity(t)','\ity^*(t)')
axis([0 20 -0.8 1.2])
```

```
%script ex8p1
%Batch least squares estimation of a single-input function
%This code is used to produce Figures 8.1 - 8.9

global c sigma

%input Gaussian mf centers
c=[0 2 4 6];

%crossing point of adjacent mfs
%cross=.9;

cross=.7;
%cross=.3;

%spread for adjacent mfs crossing at 'cross' defined above
sigma=.5*(c(2)-c(1))/sqrt(-2*log(cross));

%input data
x=0:6;

%output data
for i=1:7
    y(i)=x(i)-1*cos(1.5*x(i))+sin(0.4*x(i));
end
```

```
%construct phi for batch least squares by evaluating
%fuzzy basis functions at input data
phi=[z1(x(1)) z2(x(1)) z3(x(1)) z4(x(1));
     z1(x(2)) z2(x(2)) z3(x(2)) z4(x(2));
     z1(x(3)) z2(x(3)) z3(x(3)) z4(x(3));
     z1(x(4)) z2(x(4)) z3(x(4)) z4(x(4));
     z1(x(5)) z2(x(5)) z3(x(5)) z4(x(5));
     z1(x(6)) z2(x(6)) z3(x(6)) z4(x(6));
     z1(x(7)) z2(x(7)) z3(x(7)) z4(x(7))];

%construct Y for batch least squares
Y=y';

%calculate output mf centers via batch least squares
b=inv(phi'*phi)*phi'*Y;

%test data at which to construct characteristic
%of fuzzy system obtained by batch LS
xx=linspace(0, 6);

%evaluate nonlinear fn g at test data
g=xx-1*cos(1.5*xx)+sin(0.4*xx);

%evaluate fuzzy system at test data
%also save basis fns for plotting
for i=1:length(xx)
   f(i)=b'*[z1(xx(i)) z2(xx(i)) z3(xx(i)) z4(xx(i))]';
   zz1(i)=z1(xx(i));
   zz2(i)=z2(xx(i));
   zz3(i)=z3(xx(i));
   zz4(i)=z4(xx(i));
end

plot(xx,g,'b',xx,f,'k','linewidth',3),grid
%plot(xx,zz1,'k',xx,zz2,'k',xx,zz3,'k',xx,zz4,'k','linewidth',
3),grid
%axis([-1 7 0 1.1])
legend('\itg(x)','\itf(x)')
-----------------------------------------------------------------
function z=z1(yy)
global c sigma

z=exp(-0.5*((yy-c(1))/sigma)^2)/…
sum(exp(-0.5*((yy-c)/sigma).^2));
-----------------------------------------------------------------
function z=z2(y)
global c sigma
```

```
z=exp(-0.5*((y-c(2))/sigma)^2)/...
sum(exp(-0.5*((y-c)/sigma)0.^2));
------------------------------------------------------------
function z=z3(y)
global c sigma

z=exp(-0.5*((y-c(3))/sigma)^2)/...
sum(exp(-0.5*((y-c)/sigma).^2));
------------------------------------------------------------
function z=z4(y)
global c sigma

z=exp(-0.5*((y-c(4))/sigma)^2)/...
sum(exp(-0.5*((y-c)/sigma).^2));
```

```
%script ex8p2
%Batch LS estimation of a 2-input 1-output nonlinear
%function.
%Used for Example 8.2

clear all

num_mfs=5;              %number of rules on a universe
numrules=num_mfs^2;     %total number of rules

%input centers and spreads (fixed)
c1=linspace(-1,1,num_mfs);
c2=linspace(-1,1,num_mfs);

%calculate spreads such that adjacent Gaussians cross at
%'cross'
cross=0.5;
sigma=0.5*(c1(2)-c1(1))/sqrt(-2*log(cross));

%number of input samples
numsamplesx1=11;
numsamplesx2=11;
numsamples=numsamplesx1*numsamplesx2;

%arrays for premise membership functions evaluated at input
%data points
mmu=zeros(numsamples,numrules);
Phi=zeros(numsamples,numrules);

%array for premise membership functions
mu=zeros(1,numrules);

%training data
x1=linspace(-1,1,numsamplesx1);
x2=linspace(-1,1,numsamplesx2);
```

```
%generate input-output data pairs
k=1;
for i=1:numsamplesx2
    for j=1:numsamplesx1
        data(k,:)=[x1(j) x2(i) sin(x1(j))*(cos(x2(i)))^2];
        k=k+1;
    end
end

n=1; %rule number
for i=1:numsamples
    for j=1:num_mfs
        for k=1:num_mfs
            %premise values for rules
        mmu(i,n)=exp(-0.5*((data(i,1)-c1(j))/sigma).^2).* …
                exp(-0.5*((data(i,2)-c2(k))/sigma).^2);

            if n==numrules
                n=1;
            else
                n=n+1;
            end
        end
    end
end

for i=1:numsamples
    %construct Phi for batch LS
    %(array of fuzzy basis functions)
    Phi(i,:)=mmu(i,:)/sum(mmu(i,:));
end

Y=data(:,3); %construct Y for batch LS

%LS calculation of output singleton locations
theta=inv(Phi'*Phi)*Phi'*Y;
bb=theta;

%create input test data
xx1=linspace(-1,1,101);
xx2=linspace(-1,1,101);

for i=1:length(xx2)
    for j=1:length(xx1)
        g(i,j)=sin(xx1(j))*(cos(xx2(i)))^2;
        n=1; %rule number
        for k=1:num_mfs
            for l=1:num_mfs
                mu(n)=exp(-0.5*((xx1(j)-c1(k))/sigma).^2).*…
                        exp(-0.5*((xx2(i)-c2(l))/sigma).^2);
```

```
                n=n+1;
            end
        end
        %fuzzy system evaluated at test data
        ff(i,j)=sum(bb'.*mu)/sum(mu);
    end
end

colormap(gray)
%max(max(abs(ff-g)))
surface(xx1,xx2,abs(ff-g));
grid;
%hold on
%surface(xx1,xx2,ff)
%set(plots,'color','k','linewidth',2);
```

```
%script ex8p3
%Recursive least squares estimation of a SISO nonlinear
%function.   Used for Example 8.3

clear all

global c sigma

c=[0 2 4 6];

%crossing point of adjacent mfs
%cross=0.9;
cross=0.7;
%cross=0.3;

%spread for adjacent mfs crossing at 'cross' defined above
sigma=0.5*(c(2)-c(1))/sqrt(-2*log(cross));

numpts=201; %number of iterations in RLS algorithm

theta=zeros(4,1);   %initial parameter vector

%time history of parameter estimates saved for plotting
b=zeros(4,numpts);

%initial value of P in RLS algorithm (should be large)
P=1e5*eye(4);
```

```
for i=1:numpts
   t(i)=i-1;     %iteration time of RLS algorithm
   x(i)=6*rand; %input data
   y(i)=x(i)-cos(1.5*x(i))+sin(.4*x(i)); %output data
   phi=[z1(x(i)) z2(x(i)) z3(x(i)) z4(x(i))]'; %regressor
   K=P*phi/(1+phi'*P*phi);    %K for RLS
   P=(eye(4)-K*phi')*P;    %P for RLS
   theta=theta+K*(y(i)-phi'*theta); %parameter estimate
   b(1,i)=theta(1);    %prameter estimates saved for plotting
   b(2,i)=theta(2);
   b(3,i)=theta(3);
   b(4,i)=theta(4);
end

xx=linspace(0, 6);    %test data

%nonlinear function g evaluated at test data
g=xx-cos(1.5*xx)+sin(0.4*xx);

%fuzzy system f evaluated at test data
%z's are fuzzy basis functions
%defined in script ex8p1.m above
for i=1:length(xx)
   f(i)=b(:,numpts)'*[z1(xx(i)) z2(xx(i)) ...
                      z3(xx(i)) z4(xx(i))]';
end

%plot evolution of parameter estimates
plot(t,b,'k','linewidth',3),grid,xlabel('Time\itk'),ylabel
('Parameter Estimates')

%plot g and its fuzzy approximation f
%plot(xx,g,'k',xx,f,'k','linewidth',3),grid
%legend('\itg(k)','\itf(k)')
```

```
%script ex8p4
%gradient estimation of SISO nonlinear function
%used for Example 8.4

%clear all

%step sizes
lambda1=.1;
lambda2=.1;
lambda3=.1;

%initial guesses
cinit=6*rand(1,5);
sinit=2*rand(1,5);
binit=10*rand(1,5);
```

```
numsamples=200;
times_thru_training_data=2;
train=times_thru_training_data*numsamples;

%training data
x=6*rand(1,numsamples);
y=x-cos(1.5*x)+sin(0.4*x);
X=x;
Y=y;

for i=1:times_thru_training_data-1
    X=[X x];
    Y=[Y y];
end

%vectors for parameter estimates
c=[cinit;zeros(train-1,5)];
b=[binit;zeros(train-1,5)];
s=[sinit;zeros(train-1,5)];

%parameter updates
for i=1:train-1

    %premise values
    mu=exp(-0.5*((X(i)-c(i,:))./s(i,:)).^2);

    %crisp output
    f=sum(b(i,:).*mu)/sum(mu);

    epsilon=f-Y(i);

    %gradient update laws
    b(i+1,:)=b(i,:)-lambda1*epsilon*mu/sum(mu);
    c(i+1,:)=c(i,:)-lambda2*epsilon*((b(i,:)...
        -f)/sum(mu)).*mu.*(X(i)-c(i,:))./s(i,:).^2;
    s(i+1,:)=max(.01,s(i,:)-lambda3*epsilon*...
        ((b(i,:)-f)/sum(mu)).*mu.*((X(i)...
        -c(i,:)).^2)./s(i,:).^3);

end

%save last best parameter values
bb=b(train,:);
cc=c(train,:);
ss=s(train,:);

%test data
xx=linspace(0,6);
g=xx-cos(1.5*xx)+sin(.4*xx);
```

```
%fuzzy system output at test data
for i=1:length(xx)
   mu=exp(-0.5*((xx(i)-cc)./ss).^2);%1x5 vector
   f(i)=sum(bb.*mu)/sum(mu);%scalar
end

%plot f and g on same plot
plot(xx,g,'k',xx,f,'k','linewidth',3),grid
legend('\itg(x)','\itf(x)')
xlabel('\itx')
ylabel('\itf(x) & g(x)')

%plot evolution of parameter estimates
%plot(b,'k','linewidth',3),grid
%plot(c,'k','linewidth',3),grid
%plot(s,'k','linewidth',3),grid
```

```
%script ex8p5
%Gradient estimation of a static two-input nonlinear
%function.   Used for Example 8.5

clear all

%step sizes
lambda=.1;
lambda1=lambda;
lambda2=lambda;
lambda3=lambda;

%declare number of samples and number of rules
numsamples_onedim=11;
num_times_thru_data=200;
numsamples=numsamples_onedim^2;
numrules=4;

%input portion of training data
xx1=linspace(-1,1,numsamples_onedim);
xx2=linspace(-1,1,numsamples_onedim);

%construct input-output pairs for training data
samplenum=1;
for j=1:numsamples_onedim
    for k=1:numsamples_onedim
        x1(samplenum)=xx1(j);
        x2(samplenum)=xx2(k);
        y(samplenum)=sin(x1(samplenum))*cos(x2(samplenum))^2;
        samplenum=samplenum+1;
    end
end
```

```
%initial guesses for c's, sigma's, and a's
c1=linspace(-1,1,numrules);    %centers on x1 universe
c2=linspace(-1,1,numrules);    %centers on x2 universe

%initial spreads on x1 universe
sigma1=(.5*(c1(2)-c1(1))/...
      (sqrt(-2*log(.5))))*ones(1,numrules);

%initial spreads on x2 universe
sigma2=(.5*(c2(2)-c2(1))/...
      (sqrt(-2*log(.5))))*ones(1,numrules);

%initial consequent parameters
a0=linspace(-1,1,numrules);
a1=linspace(-1,1,numrules);
a2=linspace(-1,1,numrules);

%parameter updates for c and sigma
for samplenum=1:num_times_thru_data*numsamples

%form training data by repeating input points
%'num_times_thru_data' times
   samp(samplenum)=mod(samplenum, numsamples);
      if samp(samplenum)==0
         samp(samplenum)=numsamples;
      end

   %evaluate the approximating fuzzy system at the training
   %data based on the current values of the c's, sigma's,
   %and a's
   mu=exp(-0.5*((x1(samp(samplenum))-c1)./sigma1).^2).*...
      exp(-.5*((x2(samp(samplenum))-c2)./sigma2).^2);
   q=a0+a1*x1(samp(samplenum))+a2*x2(samp(samplenum));
   f=sum(q.*mu)/sum(mu);

   epsilon(samplenum)=f-y(samp(samplenum));

   %save current parameter values before updating
   aa0=a0;
   aa1=a1;
   aa2=a2;
   cc1=c1;
   cc2=c2;
   ssigma1=sigma1;
   ssigma2=sigma2;

   %evaluate all consequents
   qq=a0+a1*x1(samp(samplenum))+a2*x2(samp(samplenum));
```

```
    %gradient update laws
    a0=aa0-lambda1*epsilon(samplenum)*mu/sum(mu);
    a1=aa1-lambda1*epsilon(samplenum)*mu/...
        sum(mu)*x1(samp(samplenum));
    a2=aa2-lambda1*epsilon(samplenum)*mu/...
        sum(mu)*x2(samp(samplenum));
    c1=cc1-lambda2*epsilon(samplenum)*...
        ((qq-f)/sum(mu)).*mu.*(x1(samp(samplenum))...
        -cc1)./ssigma1.^2;
    c2=cc2-lambda2*epsilon(samplenum)*...
        ((qq-f)/sum(mu)).*mu.*(x2(samp(samplenum))...
        -cc2)./ssigma2.^2;
    sigma1=max(.1, ssigma1-lambda3*epsilon(samplenum)*...
            ((qq-f)/sum(mu)).*mu.*(((x1(samp(samplenum))...
            -cc1).^2)./ssigma1.^3));
    sigma2=max(.1, ssigma2-lambda3*epsilon(samplenum)*...
            ((qq-f)/sum(mu)).*mu.*(((x2(samp(samplenum))...
            -cc2).^2)./ssigma2.^3));

end

clear xx1 xx2 epsilon

%test data
xx1=linspace(-1, 1, 100);
xx2=linspace(-1, 1, 100);

samplenum=1;
for j=1:numsamples_onedim
     for k=1:numsamples_onedim
          x1(samplenum)=xx1(j);
          x2(samplenum)=xx2(k);
          y(samplenum)=sin(x1(samplenum))*...
                         cos(x2(samplenum))^2;
          samplenum=samplenum+1;
     end
end

for samplenum=1:numsamples

    %calculate premise values of all rules
    mu=exp(-0.5*((x1(samplenum)-c1)./sigma1).^2).*...
       exp(-0.5*((x2(samplenum)-c2)./sigma2).^2);

    q=a0+a1*x1(samplenum)+a2*x2(samplenum);

    f=sum(q.*mu)/sum(mu);

    epsilon(samplenum)=f-y(samplenum);

end
```

```
for i=1:length(xx2)
    for j=1:length(xx1)
      g(i,j)=sin(xx1(j))*(cos(xx2(i)))^2;
      for k=1:numrules
         mu(k)=exp(-0.5*((xx1(j)-...
               c1(k))/sigma1(k)).^2).*...
               exp(-0.5*((xx2(i)-c2(k))/sigma2(k)).^2);
      end
      %fuzzy system evaluated at test data
      q=a0+a1*xx1(j)+a2*xx2(i);
      ff(i,j)=sum(q*mu')/sum(mu);
    end
end
%colormap(gray)
%max(max(abs(ff-g)))
surface(xx1,xx2,abs(ff-g));
%plot(abs(epsilon))
```

Index

Fuzzy Control and Identification, By John H. Lilly

Printed and bound by CPI Group (UK) Ltd, Croydon, CR0 4YY
16/04/2025
14658592-0005